UFO Disclosure 2021

The Final Guide on Declassified Reports, Military Sightings, Historical Abductions and Recent Investigations About the Greatest Secret Ever: Extraterrestrial Life on Earth

Robert Davis

© **Copyright Robert Davis 2021 - All rights reserved.**

The content contained within this book may not be reproduced, duplicated or transmitted without direct written permission from the author or the publisher.

Under no circumstances will any blame or legal responsibility be held against the publisher, or author, for any Tdamages, reparation, or monetary loss due to the information contained within this book. Either directly or indirectly.

Legal Notice:

This book is copyright protected. This book is only for personal use. You cannot amend, distribute, sell, use, quote or paraphrase any part, or the content within this book, without the consent of the author or publisher.

Disclaimer Notice:

By reading this document, the reader agrees that under no circumstances is the author responsible for any losses, direct or indirect, which are incurred as a result of the use of information contained within this document, including, but not limited to, — errors, omissions, or inaccuracies.

Table of Contents

INTRODUCTION 11

    ARE WE ALONE? 11

    UFO AND ALIEN ENCOUNTERS 13

    THE ABDUCTION PHENOMENA 14

    THE BARNEY AND BETTY HILL INCIDENT 16

CHAPTER 1: UFO EVENTS IN ANCIENT TIMES 20

    UFOS SEEN IN ANCIENT EGYPT 21

    ANCIENT ROMANS 23

    THE FIRST KNOWN UFO INVESTIGATIONS 24

    WHAT WERE THESE UFOS? 25

    ANCIENT ASTRONAUT THEORIES 26

    ORIGIN AND INFLUENCE 29

    IS THERE EVIDENCE? 32

    FACTS VS. FICTION 33

    COULD THIS BE PSEUDOSCIENCE OR VEILED RACISM? 35

CHAPTER 2: THE FREEDOM OF INFORMATION ACT AND WHY IT MATTERS 40

    THE REFERENCE REPORT-PROJECT BLUE BOOK 43

    U.S. AIRFORCE FACT SHEET ON UFOS 43

CHAPTER 3: UFO EVENTS 1940S - 1950S 47

    MJ-12 Report: Majestic 12 49

    Roswell New Mexico 55

    1940: Foo Fighters 56

    Ghost Rockets – Scandinavia 57

    Washington DC UFO incident 59

    Fiorentina Stadium UFO incident 61

    Project Blue Book 62

CHAPTER 4: UFO EVENTS OF THE 70S, 80S, AND 90S 64

    Tehran, Iran UFO Incident 64

    The Cerenchovo (Siberian Russia) UFO Incident 66

    Bob Lazar and Area 51 70

    The Public Revelation 71

    Lazar's Claims and Evidence 72

    Element 115 72

    Lazar's Story and Shortcomings 73

    Area 51 74

    The Secret Base 74

    Where is the Truth? 76

    1994 Ruwa, Zimbabwe Incident 77

CHAPTER 5: THE DISCLOSURE PROJECT 83
GOVERNMENT COVER-UPS 86

CHAPTER 6: RECENT U.F.O. SIGHTINGS 89
NIMITZ CARRIER STRIKE GROUP-2004 89
A PUZZLING PRESENCE AT 80,000 FEET 90
NIMITZ INCIDENT PROLONGED ATTENTION 92
O'HARE INTERNATIONAL AIRPORT SIGHTING 95
U.S.S. THEODORE ROOSEVELT U.F.O. SIGHTING 98

CHAPTER 7: RECENT UFO INVESTIGATIONS AND DEVELOPMENTS 102
TO THE STARS ACADEMY 102
THE PENTAGON'S NEW U.F.O. TASK FORCE 106
SHOULD D.O.D. BE CONCERNED? 108
ARE REPORTS OPEN-ENDED? 109
ASTEROID OUMUAMUA 111

CHAPTER 8: THE JUNE 2021 REPORT 113
THE JUNE 2021 REPORT 114
WHAT WILL THE REPORT FIND? 115
THE PENTAGON AND U.F.O.S 117
GROWING PUBLIC BELIEF 119

CONCLUSION 122

REFERENCES 127

UFO REPORT UPDATE – JULY 2021 128

    BACKGROUND 128

    WHAT THE REPORT SAYS 130

    CONCLUSIONS AND QUESTIONS 138

    CLASSIFIED VERSION 143

    SUMMARY 145

Table of Abbreviations:

UFO- Unidentified Flying Objects

UAP- Unidentified Aerial Phenomena

AAV- Anomalous Aerial Vehicles

FAA- Federation Aviation Administration

USAF- United States Air Force

AATIP- Advanced Aerial Threat Identification Program

MOA- Military Operations Area

NASA- National Aeronautics and Space Administration

UAPTF- Unidentified Aerial Phenomena Task Force

NARA- National Archives and Research Administration

FOIA- Freedom of Information Act

FIO- Freedom of Information Office

ATIC- Air Technical Intelligence Centre

FOGI- Foreign Operations and Government Subcommittees

USSR- Union of Soviet Socialists Republic

AA- Ancient Astronauts

CIA- Central Intelligence Agency

FBI- Federal Bureau Investigation

NM- New Mexico

BBC- British Broadcasting Corporation

MJ-12- Majestic 12

Introduction

As the name suggests, Unidentified Flying Object, or UFO, refers to any object in the sky of unknown origin. The U.S. military calls them Unidentified Aerial Phenomena (UAP), or Unexplained Aerial Phenomena.

Throughout history, UFO sightings have been reported in various parts of the world, raising questions and concerns about life on other planets and whether extraterrestrials have visited Earth.

After World War II, UFOs became a significant subject of interest and the inspiration behind numerous films and books following the development of rocketry.

Are We Alone?

"Are we alone on the planet Earth? Do aliens exist, and are UFOs real?"

This question often becomes a theme in Hollywood films and science fiction publications. However, these concerns may be less debatable today than they used to be.

On 17th December 2017, the U.S. Department of Defense (the Pentagon) released three classified videos of events that occurred in 2004 and 2015, which they referred to as "unexplained aerial phenomenon," which had U.S. Naval pilots bumping into UFOs. This got UFO space enthusiasts and science fiction writers to revise their theories on other life forms in the universe.

These were not new videos, and two of them had been published by New York Times in 2017. The State Department of Defense confirmed their authenticity in September 2019 and released these three declassified videos after a "thorough review" and bid "to clear up any misconceptions by the public on whether or not the footage that had been circulating was real or whether or not there is more to the videos."

Following thorough reviews, the then pentagon Spokesperson Sue Gough issued a statement saying, "the department has determined that the authorized release of these unclassified videos does not reveal any sensitive capabilities or systems and does not impinge on any subsequent investigations of military air space incursions by unidentified aerial phenomena."

Also referred to as 'Flying Saucers,' UFOs have been reported throughout the twentieth century.

Science fiction writers have written and published about alien life forms visiting Earth in flying saucers. On the other hand, conspiracy theorists have raised accusations of cover-ups by governments about the existence of UFOs.

Studies over decades have established that most UFO sightings are a case of misidentification, while some are hoaxes, and a very small percentage are left unexplained.

In 2007, the Pentagon launched a program that studied recordings of such aerial encounters. This program ended in 2012 in what was referred to as "incapability for high enough priority for funding." Luiz Elizondo, the former head of the program, told CNN in 2017 that there exists evidence that the planet is being visited by extraterrestrials and that we might not be alone.

The release by the Department of Defense is a step towards bringing greater transparency in official research into alien life forms and UFOs and add legitimacy to theories about the presence of extraterrestrial lifeforms.

UFO and Alien Encounters

Although he passed away more than half a century ago, George Adamski remains one of the most curious and controversial characters in UFO history. Some thought of him as a prophet, and others, a laughingstock that sought publicity. In the late 1940s, Adamski claimed UFO fame after he took photos of what he insisted were flying saucers.

Adamski took countless photos from the 40s, which he insisted were UFOs, or as he called them, flying saucers. However, J. Allen Hynek, who was the scientific consultant to the Air Force's Cold War-era UFO investigation team Project Blue Book, and other experts, dismissed them as fakes.

In 1952, Adamski resurfaced with new claims that he had met and conversed with a visitor from Venus in a California desert through gestures and telepathy, raising an even stranger story. He documented his adventures in books, the first being *Flying Saucers Have Landed* (1953), which was co-authored by Desmond Leslie, who narrated his conversation with the Venusian. This book was largely read at the time and gained a new generation of fans in the 1960s.

In 1955, he published another volume, *Inside the Space Ships*, which narrated further encounters with envoys from Mars and Saturn. His narrations depicted that every planet in the solar system is populated with human-like inhabitants, just as the dark side of the Earth's Moon.

His claims in the second book shared how he was taken aboard one military ship by his new friends and flew to an enormous mother ship that hovered over the Earth. Allegedly, his new friends gave him a ride over the Moon.

As shocking as his stories were, Adamski's fame rose as he became an international celebrity who would lecture widely on the gospel of UFO. In 1959, Adamski was invited by the queen of the Netherlands to her palace to discuss extraterrestrial activity, and later, in 1963, he secretly had a meeting with the pope. Later, Adamski had gained a massive following all over the world, apart from a few, including Arthur C. Clarke, the author of *2001: A Space Odyssey*, who denounced Adamski's work and his believers as "nitwits."

The Abduction Phenomena

These autobiographical memories are often doubtful. Surprisingly, a large number of people have been reported to have been abducted by extraterrestrials. These accounts are hard to dismiss as merely deception or insanity.

There are many reasons that may prompt the doubt existence the existence of extraterrestrials, but thousands and perhaps even millions of individuals report memories of abductions.

The explanations for this is hotly debated and some think they are wild fantasies. With the aid of hypnosis, many individuals have found themselves in somnambulistic state talking about an experience that their conscious mind has no recollection of.

Despite there being no universal agreement regarding the contextual or experiential features that are necessary to define alien abduction experiences, several descriptions have generally been proposed. The most common are experiences of abduction and examination.

The self-reported abductees are a part and parcel of community members who participate in cultural myths that can directly be traced to Barney and Betty Hill and whose stories came about after aliens started abducting people in TV plot lines and series in the 1950s. Aliens were installations of science fiction in the 20th century after reports of flying saucers came into popularity in the 1940s.

The English professor, Terry Matheson, in his 1998 book, *Alien Abductions: Creating a Modern Phenomenon*, says, "We in the late 20th century may well be in the exciting position of being able to observe and study myth in the process of being created."

In a report published by the Washington Post in 2013, 2.5% of the U.S. population has reported having had some personal experience with alien abductions.

Studies by nun-ufologists have, however, identified truths that apply to these categories of people, including the Hills. They indicated that these people, at least the majority, were not lying and that although their arguments of being kidnapped by aliens are not supported by a shred of evidence, they were, indeed, an experiencer of some deeply impactful event. Their findings indicated that these people were not crazy, at least not in the way we think when the topic of crazy people surfaces, although they differ from the rest of the population on some key psychological traits.

The Barney and Betty Hill Incident

One night, Barney and Betty Hill were heading to their rural home in Portsmouth, New Hampshire, driving on a rural highway. The date was 19th September 1961. They were an ideal couple. Barney worked as a postman while Betty was a social worker. Both were active in their community and the civil rights movement.

The couple recounted seeing a bright object that appeared to be following their car on an isolated road that curved through the White Mountains and later got home at around 5 in the morning feeling that something dreadful had happened to them and unable to account for two hours of their night.

It took them a while, and eventually, they were able to remember the life-changing event that came upon them on the lonely stretch of the highway.

They had been kidnapped by aliens.

Their claims became the first publicized account of alien abduction, which was first told by them to a psychiatrist and later in a book and T.V. series, forming a template for countless abduction stories.

Though details of the abduction stories vary, many accounts follow the basic script of the Hills' self-reported encounter, which explained how they were taken by other beings and subjected to various experiments and later returned, never to be the same.

The Abduction

In their book, *The Interrupted Journey* (1996), co-authored with John G. Fuller, the Hills recount the cloudless and starry night when they stopped for coffee at a roadside dinner at around 10 pm and figured they would get home at around 3 am.

The couple got back in the car and continued with their journey home. Along the way, Betty looked out the passenger side window and saw a bright object following them. She insisted to Barney that they park the car and followed to look at the object through a pair of binoculars.

This is when Barney saw an object like a spacecraft with a set of double windows through which he could see "dozens of living beings" with uniforms and who looked straight at him.

Terrified, he rushed back to the car and pulled off the highway onto a winding side road to lose the spacecraft. This was followed by a series of loud beeps. Both then felt "odd tingling drowsiness come over them," as written by Fuller, and a "sort of haze."

When they regained consciousness, the couple was 35 miles down the highway heading home.

In the beginning, few oddities were noted. They both were disheveled, and Barney's shoes were scuffed. The binoculars strap was also broken.

Barney felt like something had happened to his body and went to examine himself in the bathroom. Days later, Betty began having nightmares. Barney told a friend that his feeling was "one of a person who saw something he doesn't want to remember."

After two years since their abduction, the couple visited a psychiatrist. Under hypnosis, both recounted what had occurred the night at the White Mountain.

They referred to seeing short, gray aliens with big, wraparound eyes, who took them aboard their saucer-shaped craft and probed them with needles.

The psychiatrist didn't believe their story, but he could see that the Hills did.

The recollection terrified Barney. On the other hand, Betty got comfortable sharing the story with friends and others and extensively gave lectures about the experience at schools and to local groups.

Journalists started sniffing around. Soon, the Hills agreed to a book deal. NBC aired The UFO Incident in 1975, which led to the creation of the T.V. movie based on the Hills' story.

After two years since it was aired, reports of abductions, something that had never been recorded or heard of before Betty and Barney Hill, rose by 2,500%.

Unlike the thousands of abductees who followed, the Hills remembered or believed to remember their experiences. The majority who believed to have been abducted didn't recall the particulars and only relied on the strong sense that they had been abducted.

In her book; *Abducted: How People Come to Believe They Were Kidnapped by Aliens* (2005), psychologist Susan Clancy quotes a 22-year-old man called Tom: "About a month ago, I got out of the shower and looked at my back and saw bruises, then I talked to my mum about it, and she looked at my back and said, 'well, maybe aliens.'"

In such an instance, Tom would be a typical abductee, for something unusual occurred, and while other possibilities are considered, Tom's friend suggests that he might have gotten the bruises at his job on a demolition site.

"There is a tendency to either be uncritical or label the narrators as crazy when hearing abductee stories. That's unfair and not supported by the evidence," wrote Christopher French, a psychology professor and head of anomalistic psychology at Goldsmith.

Psychopathy, according to French notes, is not more common among abductees than among the general population.

Chapter 1: UFO Events in Ancient Times

140,947 events of UFO sightings dating back to the 20th century have been reported in the last 43 years since the National UFO Reporting Centre began keeping track (Monfort, 2017).

The majority of these sightings have been recorded in the United States. Canada, and Australia. Nordic countries have also recorded significant numbers of UFO sightings. While different research analyses have not exactly recorded what they are, the majority have indicated their suspicions as either airplanes, satellites, comets, or fireworks, thus expanding the doubts of the many people who write off UFO spotters as fanatics.

UFOs Seen in Ancient Egypt

1. A plane glides over an ancient spider geoglyph in the Peruvian desert. Source; NAT GEO Collection by Robert Clark

The phenomenon of UFOs stretches further back in history than 1950s' flying saucer reports.

History records the oldest sighting of an unidentified flying object as far back as 1440 BC. The Royal scribe of an Egyptian Pharaoh documented the incident.

Before the modern era, which can be readily explained as fireworks and airplanes, UFOs were difficult to trace and record, and their records were stored by the ancient Greeks, Romans, Indians, Chinese, Japanese, Mexicans, and more.

"In the year 22, of the 3rd month of winter, the sixth hour of the day... the scribes of the House of Life found it was a circle of fire that was coming in the sky... It had no head; the breath of its mouth had a foul odor. Its body one rod long and one rod wide. It had no voice. Their hearts became confused through it; then they laid themselves on their bellies...They went to the Pharaoh... to report it. His Majesty ordered.... [An examination of] all which is written in the papyrus rolls of the House of Life. His Majesty was meditating upon what happened. Now after some days had passed, these things became more numerous in the sky than ever. They shone more in the sky than the brightness of the Sun and extended to the limits of the four supports of the heavens... Powerful was the position of the fire circles. The army of the Pharaoh looked on with him in their midst. It was after supper. Thereupon, these fire circles ascended higher in the sky towards the south... The Pharaoh caused incense to be brought to make peace on the hearth... And what happened was ordered by the Pharaoh to be written in the annals of the House of Life... so that it be remembered forever." (Fontaine, 1985)"

This quote comes from the Tulli Papyrus which is now recorded to have disappeared.

After the director of the Vatican Museums, Alberto Tulli, found the document while in Cairo, plenty of controversies surrounded it, not the least because all that remains are copies, the original being lost to the ages. The Vatican never received this papyrus. Rather Tulli kept it in his collection, which was passed down to his descendants and subsequently lost.

The papyrus opens with "in the year 22," which in the Gregorian calendar equates to the year 1440 BC. This would mean that the Pharaoh mentioned (but never named) would be Thutmose III, who reigned from 1504 to 1450 BC.

In spite of this, the 'circles of fire' exact meaning as described is hard to determine. This is due to the existence of only one copy of the original. Confirming whether there are any typos in transcription or translation cannot be confirmed. There are no other sources to back up the Tulli Papyrus claim due to the limited number of written works from that age.

Ancient Romans

The Romans are the ones who accumulated several reported sightings from whatever form of UFOs have ever been recorded. Historians such as Pliny the Elder, Livy, and Plutarch are the first men to make these sightings.

Their sightings are widely regarded as the most accurate and reputable due to the rigorous procedures the Roman authorities demanded before any event could be recorded in the official annals. That said, it is worth noting that the incidences could be talking about meteorites or comets, which to the ancient eyes would have seemed otherworldly.

Below are examples of ancient Roman UFO sightings:

a) "Navium gleamed the sky (ships)." In 218 BC
b) In 217 BC, "at Arpi parmas were seen in the sky (round shields)."
c) In 212 BC, "at Reate saxum was seen flying (huge stone)."
d) In 173 BC, "at Lanuvium, a fleet was said to have been seen in the sky."
e) In 154 BC, "at arma appeared flying in the sky (Compsa weapons)."
f) In 104 BC, "Ameria and Tuder saw weapons in the sky rushing together from east and west, those from the west being routed."
g) In 100 BC, in Rome, "a clipeus, burning and emitting sparks, ran across the sky from west to east, at sunset. (round shield)"
h) In 43 B.C., in Rome, "armorum telorumque left earth to the sky with noises (defensive and offensive weapons)."

The First Known UFO Investigations

Throughout much of antiquity, records on UFOs were hardly kept.

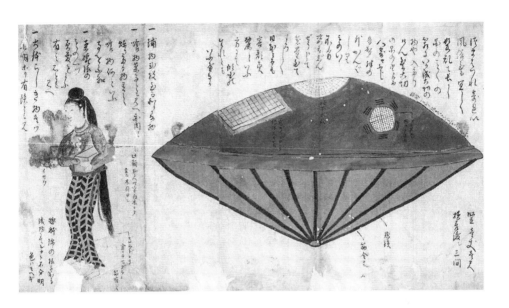

2. Courtesy National Archives of Japan

The first known and documented investigation into a possible alien presence took place in Japan in 1235.

One night, a high officer named General Yoritsume and his army were settling down in their camp when they spotted mysterious lights in the sky. The general and his troops watched in astonishment as these lights performed amazing acrobatic movements, such as circling endlessly and flying in loops. Baffled by the bizarre aerial display, General Yoritsume ordered a scientific investigation of what he had just witnessed. The explanation Yoritsume's scientists gave the general oozed with comfort and calm. "The whole thing is completely natural," Yoritsume was told about the mystery lights. "It is only the wind making the stars sway." (Maloney, 2011)

What Were These UFOs?

In recent history, we are left with the descriptions given by ancient historians. These, however, are undoubtedly genuine accounts of things people witnessed, but what exactly they saw may never be known for certain.

The certainty that these events occurred cannot be proven as to whether they are natural or unnatural. NASA scientist Richard Stothers concludes in his report of classical UFO claims that:

"It is accordingly impossible to know whether the later observers (mostly practical Romans) interpreted the phenomena literally as they described them or were simply using the best descriptive language they were capable of while holding back on theoretical speculation. But any viable theory must reckon with the extraordinary persistence and consistency of the phenomena discussed here over many centuries. Whether one prefers to think in terms of universal recurrent visions from the collective unconscious, misperceptions of ordinary objects, unusual atmospheric effects, unknown physical phenomena, or extraterrestrial visitations, what we today would call UFOs to possess and the increase of human knowledge." (Stothers, 2007)

Ancient Astronaut Theories

Ancient astronaut theories have made their way to fame in history books for some time. However, this does not mean they are true.

In 1966, a book titled Intelligent Life in the Universe was released by science communicator Carl Sagan and his Soviet colleague Iosif Shklovsky. This was an English-language version of Shklovksy's *Universe, Life, Intelligence* that was published in 1962 in the Union of

Soviet Socialists Republic (USSR). This book remains one of the most highly influential books on the topic of extraterrestrial intelligence and the search for its existence.

In Chapter 33, "Possible Consequences of Direct Contact," the authors claimed that scientists studying UFOs, whether supporting or rejecting their actuality, should accept the possibility that extraterrestrials have visited Earth in the past.

They further suggest that evidence of past contact could be preserved in the folklore and mythological traditions of various cultures.

Sagan and Shklovsky make reference to claims of contact made between the people of the Pacific Northwest, referred to as the Tlingit and the French expedition in 1786, La Perouse.

A century later, this event was recorded by anthropologist G.T. Emmons during one of his many visits to the area.

The descriptions of these accounts proved that it was a faithful retelling of an actual event, though it was told in the way and time of Tlingit mythology and contained supernatural elements such as the sailing ships, which were described as immense black birds with white wings.

Sagan and Shlovksy republished the narrative, indicating that it could be true that extraterrestrials visited Earth.

The meeting of La Perouse and the Tlingit lends credence to the idea that brief contact with an extraterrestrial people can be saved in a reconstructible way. The reconstruction will be increased in efficacy if:

1. It is written down very soon after the occurrence.

2. The encounter causes major changes in the society.
3. The contacting civilization does not attempt to hide its foreign nature.

Shklovskii and Sagan quoted the Sumerian legend of Oannes (Uanna to the Akkadians), ancient beings that were said to have taught agriculture, mathematics, and the arts to early humanity and had the look of fish.

They pointed to this as a possible instance of paleo-contact because of the level of consistency and detail involved. They spotlighted the need for incredulity.

They further stated that the ideas were conjectural and unfounded. This was in large part due to the unproven theories that gathered fame by the end of the 1970s. They came to "Ancient Astronaut" theories, which are less popular in the modern world.

To simplify this, the Ancient Astronauts (AA) theory confirms that extraterrestrial intelligence visited Earth in the past and made contact with ancient humans; "paleo-contact."

In some cases, exponents of UFO maintain that contact played a role in our technological, cultural, or spiritual evolution in what is widely known as the "culturalist" school of thought.

Some "culturists" argue that aliens were responsible and assisted in constructing the ancient monuments and structures.

In other cases, it is argued that these monuments paid homage to alien species, whom ancient people viewed as gods. This version of the ancient astronauts' hypothesis has become rather popular, as illustrated by the History Channel series Ancient Aliens.

Meanwhile, other advocates of the ancient astronauts' theory insist that humans have aliens as their descendants whose DNA of ancient primates was altered to foster intelligence in hominids, what could be called "geneticists."

The subscribers to this school of thought argued modern-day humans are the results of genetic alteration.

Alterations of this argument include speculations that aliens contributed their genetic material to hasten human evolution and that they (humans) are descendants of alien colonists.

Origin and Influence

The A.A. theory gained interest in the mid-20th century as an actual theory, although its existence back-dates to the late 19th century as examples of paleo-contact can be found in science fiction.

In 1954, British pseudo-historian Harold T. Wilkins proposed that UFO abductions have occurred since ancient times.

The Morning of the Magicians was released and published in 1960 by Louis Pauwels and Jacques Bergier, who were French journalists and conspiracy theorists. This merged elements of pseudohistory and occultism with theories about UFOs, conspiracies, and ancient astronauts.

Robert Charroux, a French science-fiction author, wrote about ancient astronauts in his book, *One Hundred Thousand Years of Man's Unknown History*, along with UFOs and apocalyptic prophecies.

The well-known author Erich Von Däniken, a Swiss, profiled for his fraud, popularized the A.A. theory with his 1968 book, *Chariot of the Gods*. In this and subsequent books, Von Däniken established arguments cited by A.A. theorists to this day.

He made many arguments, amongst them the claim that extraterrestrials were responsible for the greatest technological innovations in the ancient world, whose evidence can be found in ancient structures and artifacts.

In particular, Von Däniken claimed that the structures were constructed through means and designs whose technology was not available to the indigenous culture at the time. Examples of these structures are the pyramids of Giza, the Moal of Rapa Nui, the Nazca lines of Peru, Pumapunku, and the "Baghdad batteries."

In addition, the ancient artwork and iconography show spacecraft, extraterrestrials, and advanced technology, as Von Däniken would argue. These include "helicopter hieroglyphs," the Japanese Dogū figurines that he claimed resembled astronauts, in Seti at Abydos, the Egyptian Temple. Another cited evidence was the way disparate cultures had common elements in their artwork.

Finally, Von Däniken claimed that many ancient religions like the Egyptians, Mesopotamians, Hinduism, etc., were inspired by paleo-contact. He referred specifically to instances where ancient people refer to "chariots in the sky" and other divine revelations were instances where humans communicated with aliens.

His book would prove to be influential and would inspire several similar books that were published throughout the 1970s despite being panned by academics. These included *The 12th Planet* (1976), by Zecharia Sitchin, whose argument cited that ancient Sumerians were descendants of aliens from a planet called Niburu.

He claimed that this planet orbits the Sun at a great distance and takes 3,600 years to complete a single orbital cycle. Sitchin based this on his definitions of Sumerian and Middle Eastern texts, which he claims in the book mention a "12th planet" associated with the Babylonian god Marduk translated to later being associated with Jupiter.

In 1976, Robert K. G. Temple released his book, The Sirius Mystery. He argued that the oral traditions of the Dogon people of West Africa included a description of paleo-contact.

Temple argued that contact took place between the locals and an advanced species from Sirius roughly 5000 years ago. These alien species were credited by the Dogon due to their inspiring myths of Mesopotamia, Egypt, Greece, and other cultures.

The theories remained popular through the counter-cultural groups during the 1970s and were championed as part of news telepathy and fascination with primordial cultures.

By the early 1980s, interest in the ancient astronauts' theory began to diminish. In recent decades, it has experienced something reawakening.

Is There Evidence?

The supporters of the 'ancient astronauts' theory rely on a combination of oral traditions, literary references, and scientific research on archaeology, culture similarities, and gaps in historical and archaeological records, which in turn, support their arguments.

Von Däniken's biggest claim, for instance, is that ancient cultures did not have the sophisticated technology needed to create structures and artifacts. Other arguable "evidence" are ancient texts that refer to the gods, which Von Däniken claims were aliens.

In *Chariot of the Gods*, Von Däniken cites paragraphs from Sanskrit's epic Ramayana, where the gods travel to and fro on flying chariots, which are referred to as "Vimana." This, he claims, was a reference to spacecraft.

Meanwhile, Sitchin's theories are based in their entirety on the reading of the Babylonian creation myth, Enûma Eliš. He specifically claims that the Annunaki, which are the most powerful and important gods in the Babylonian sanctuary, were a technologically advanced race of humanoids from the planet Niburu.

Sitchin argues that Niburu is an undiscovered planet beyond Neptune that has a highly concise orbit which causes it to enter the Solar System once every 3,600 years. This is the "12th planet" in Sumerian mythology, for they counted the Sun and Moon as planets alongside Pluto, which was also included in the list.

Further, he claims that the 12th planet, Niburu, whose name was later changed to Marduk, collided with the Asteroid Belt, another planet located between Mars and Jupiter. This planet was known as Tiamat. This is based on the historical tale of a goddess overthrown by Marduk, the debris of which formed planet Earth, in the creation myth.

Sichin claims that Tiamat first arrived on Earth probably 450,000 years ago, looking for minerals like gold, which they grew tired of mining. This resulted in them creating a slave race, in this case, humans, by combining their DNA with that of Homo erectus. He further claimed that these accounts coincide with biblical texts.

As for Temple, there exist similarities between the folklores of the prehistoric Sumerians, Egyptians, Greeks, and the Dogon people. These include advanced astronomical knowledge, which he claims is evidence of paleo-contact as well as shared myths, symbols, and ideas.

Other supporters like a former sports commentator and current consulting producer on Ancient Aliens, Giorgio Tsoukalos, even progress that extraterrestrials were responsible for causing the Cretaceous–Paleogene extinction event, which resulted in the deaths of dinosaurs, as part of an attempt at directing evolution.

Facts vs. Fiction

Should this sound familiar, it's most likely because the idea is quite popular in science fiction circles. The ordinary example is the 1968 film, *2001: A Space Odyssey*, a film that depicts a highly advanced ETI coming to Earth in the past and the human attempts to meet up with them in the upcoming. This film was the work of writer and director Stanley Kubrik and science communicator Arthur C. Clarke.

The film was also inspired by two short stories written by Clarke: "Encounter in the Dawn" (1953) and "The Sentinel" (1951). These accounts were the basis for Parts I and II of the film. Part one is scripted to have taken place millions of years ago and involves extraterrestrials visiting Earth at a key point in humanity's evolution.

The latter part takes place around the turn of the century when humanity uncovers an alien artifact on the surface of the Moon.

More examples include the recent release of Star Trek (S6E20, "The Chase"). This episode features the crew discovering that an ancient antecedent race was responsible for the creation of life in their quadrant.

Fans of Stargate instantly recognize how this series was made based on the basis of ancient astronauts' conjecture. The story revolves around how paleo-contact between early humans and Goa'uld, the ancient species, resulted in the generation of Egyptian civilization and how their gods were the Goa'uld.

Another recent example is *Prometheus*, the 2012 film by Ridley Scott that was inspired by *2001: A Space Odyssey*. An ancient race of engineers or "Space Jockeys" from Alien is featured as responsible for creating humanity, as well as the xenomorphic that is the franchise's main antagonist.

It is a fascinating concept that touches on multiple scientific theories and pursuits, and not only the search for extraterrestrial intelligence and abiogenesis, which is the origins of life. But in the hands of people like Von Däniken, and other theorists, this quickly became recognized as something between new-age spiritualism and pseudoscience. As a result, the acceptance Von Däniken's book got from the academic community has been resoundingly negative.

Could This Be Pseudoscience or Veiled Racism?

People gathered and in large voices offered harsh critiques about his theories, his research methodology, and his conclusions: you would doubt a shortage for that.

Sagan and Shklovksii were two such individuals. Sagan wrote in his book, Broca's Brain, that he and Shklovskii feel apologetic for whatever role they played in the 1970s ancient astronaut mania.

Sagan referred to Von Däniken "and other uncritical writers" who offered their theories up as unsubstantiated proof rather than food for thought.

At face value, A.A. theories are easy to debunk because of their sloppy research, falsifiable premises, confirmation bias, lack of evidence, and disprovable claims.

The main reason among these involved the claim that ancient people were not capable of creating ancient monuments and megalithic structures.

In challenging this, archaeologists produced evidence that demonstrates the techniques that Egyptians, Mesoamericans, Mesopotamians, Incas, and other cultures used.

In either case, what mattered was having the right materials, basic tools, a comprehension of engineering and geometry, and thousands of laborers, all of which these empires possessed. Further, academics have debunked ancient astronaut's claims that megalithic structures appeared suddenly by pointing to the archaeological and historical records.

An instance of this is in Peru. Archaeologists discovered evidence of rope-and-lever systems which were used to transport stones to build the fortress of Sacsayhuamán by the Inca, north of their ancient capital of Cusco.

Incan Nazca Lines were made similarly to how other geoglyphs around the universe were. It involved scraping the top layer of the ground to expose brighter patches underneath. They were easy to plot out using simple measurements and a hillside viewing as well.

As for the Egyptian Pyramids, archaeologists have been able to construct detailed accounts of how they were constructed. This has included data of the location of the quarries, where the stone came from, the tools they used, the engineering techniques they involved, and the written records that describe their purpose and the organization of the labor.

The "gaps" in the record that ancient astronauts theorists refer to are gaps in their apprehension. This is a classic instance of the 'appeal to ignorance' delusion, where a supposed lack of evidence to the contrary is taken as proof.

The other common logical fallacy is the analogy known as "Russell's teapot." Many ancient astronaut theorists have argued that their ideas cannot be disproven whenever they are criticized for not meeting the burden of proof.

The other piteous reproval of ancient astronaut's theories involves the way they resemble racist and ethnocentric theories on artificial sites throughout the universe.

Between the 16th and 19th centuries, European travelers observed several monumental structures in Africa, Asia, and the Americas for the first time and misreported their origins. Often, they concluded that the local population was not capable of creating such feats of construction and engineering.

A good example includes Angkor Wat, ancient structures in modern-day Cambodia that covers an area of about 155 mi^2, equivalent to 400 km^2. The ruins were once the capital of the Khmer Empire, which sprawled across the better part of South-East Asia.

However, when the temples were first observed by a French naturalist in 1860, he theorized that they ought to have been constructed by the ancient Romans. He also thought of Alexander the Great because he assumed the indigenous Khmer were incapable.

Comparably, there are the large rocky mountains at Cahokia, Illinois, the largest of which is known as Monks Mound, built in the 1050 Common Era by the Mississippians. Before the city went down by 1400

CE, it had a population of 25,000 to 50,000 people who grew maize, hunted with fine arrow points, and created pottery, shell jewelry, and flint clay figurines.

Soon, by the 17th century and after, white missionaries and settlers began moving to the area and were denied the responsibility of the local people.

For some, they went as far as claiming that it was the Egyptian travelers that were responsible. Relative claims about the stepped pyramids of Mesoamerica, such as the Chichen Itza, Teotihuacan, have been made.

Again, the thoughts that traveling Egyptians designed or built these structures and not the Mayans, the Toltecs, Aztecs, and other local people, remain a popular one even to this day even though classicists have thoroughly confuted them using archaeological evidence, carbon dating, and other methods.

Both the Egyptian and American pyramids were built thousands of years apart. This can be seen through carbon dating and archaeological evidence, which show that different techniques were used when they were built.

Other demonstrations show that they served different purposes- where the great Pyramids were tombs for the Pharaohs while Mesoamerican stepped pyramids were linked to religious rites. Except for the Viking expeditions, no evidence exists of trans-Atlantic contact before the "Columbian Era," late 15th/early 16th centuries.

Ancient astronauts' theories narrow to the Eurocentric belief that only cultures from the Mediterranean or West Asia were capable of megalithic engineering and would be where modern western history traces its roots.

In respect to this, ancient astronauts' theories are almost similar, only that aliens have become a substitute for European or Mediterranean civilizations. In the end, it is assumed, without evidence, which indigenous human beings weren't up to the task.

The other reason that ancient astronauts' theories remain popular is that they appeal to a wish-fulfillment sense. The idea that new extraterrestrial sophistication exists out there and has visited Earth before is intriguing and fascinating in the end.

Hence, it may not be all about racism and conspiracy theories because it doesn't change the reality that these beliefs may be built on the disapproval of the scientific method.

The belief denigrates our ancestors by denying their accomplishments. Believing and accepting the existence of intelligent life elsewhere in the cosmos is a positive thing. However, it should not cause us to lose sight of our fellow humans and the irrefutable spark of creativity we all possess.

Chapter 2: The Freedom of Information Act and Why It Matters

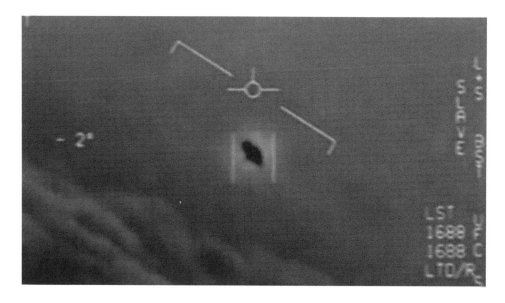

3. File photo of a UFO released by the U.S. government recently.

The existence of UFOs may seem like the domain of science fiction. However, Representative John Moss of California did not discriminate in his pursuit to open as much government information as possible to the public when he laid the foundation for legislation that led to FOIA in 1966. As Congress held hearings to debate the purview of Moss's legislation during the 1950s and 1960s, the committees chaired by Moss addressed the cunning problem as a 'simple problem.'

These committees included the Special Government Information Subcommittee and the Foreign Operations and Government Subcommittee (FOGI) of the Committee on Government Operations.

Every year, vast amounts of information were produced by the federal government. But of that stack of data, the subcommittee needed to know what the government could release, as well as what federal officials should restrict.

For more than three decades, government UFO records have been sampled and uploaded for download and access by different websites, thanks to the efforts of some intrepid truth-seekers.

The massive data collection includes more than 2500 pages of Unidentified Flying Objects, also referred to as the "Unidentified Aerial Phenomena" (UAP) by the U.S. government; related documents have been declassified by the CIA from the 1980s.

The Black Vault, an online repository for UFO-related documents ran by John Greenwald Jr., an author, contains documents that were obtained through strings of the Freedom of Information Act (FOIA) requests which were filed over the last quarter-century. Over the years, more requests were filed and piled up.

The CIA created a CD-ROM full of declassified documents which they referred to as "The UFO Collection." In mid-2020, Author Greenwald purchased the CD-ROM and uploaded most of its contents as a series of searchable pdf files on his website.

These uploaded documents cover incidents dating back and including the 1976 account of the government's then Assistant Deputy Director for Science and Technology. The Deputy was hand-delivered a mysterious piece of intelligence On a UFO that had a detailed description of a mysterious midnight explosion in a small Russian town.

According to Greenwald, as published on his website, he stated, "Although the CIA claims this is their entire declassified collection, there may be no way to verify that entirely. Research by the Black Vault will continue to see if there are additional documents still uncovered within the CIA's holdings."

The U.S. Air Force retained custody of the National Archives for the records on Project BLUE BOOK relating to the investigations of UFOs.

Project Blue Book has been averted and its records made available for examination in multiple research rooms owned by the U.S. Air force, who have retained their defense that since the closure of the project in 1969, they have no information on sightings after that date.

The National Archives has also received numerous inquiries concerning documents such as "MJ12" and "Briefing Document: Operation Majestic 12." The Truman and Eisenhower Libraries have searched their holdings for any references to, or copies of, UFOs documents. The records of the NSC for the Truman and Eisenhower Administrations have since been put under the custody of the National Archives.

Digging for documents related to UFOs was previously designed with indexes to the NSC's Policy Paper and Meeting Minute files under the subjects 'MJ-12, majestic, unidentified flying objects, UFO, flying saucers, extraterrestrial biological entities, and Aquarius.'

These searches were all negative except for the one: "Memorandum for General Twining, from Robert Cutler, Special Assistant to the President, Subject: "NCS/MJ-12 Special Studies Project," which was dated 14th July 1954.

The Reference Report-Project BLUE BOOK

Text records of Project Blue Book, which excludes the names of people involved in the sightings, have been made available for research in the National Archives Building.

The records include about two cubic feet of the jumbled project or administrative files, 37 cubic feet of case files, with each sighting arranged in chronological order, and three cubic feet of documentation associated with the Office of Special Investigations (OSI). The portions of these are arranged chronologically, by OSI district, and by overseas command.

A cubic foot of information comprises about 2,000 pages which can be accessed from the National Archives Microfilm reading room, and which also has access to the Blue Book textual records is through 94 rolls of 35mm microfilm. All the rolls and the finding aids are contained in the first microfilm roll of contents.

Photographs distributed in the textual records have been filmed separately on the last disks, rolls. Motion and picture films, sound recordings, and still pictures are retained by NNSM and NNSP.

U.S. Airforce Fact Sheet on UFOs

In January 1985, Wright Patterson distributed sheets of a copy of the U.S. Airforce findings. It read:

The United States Air Force
Public Affairs Division,

Wright-Patterson AFB,
Ohio 45433

UFOs & PROJECT BLUE BOOK

On 17th December 1969, the Secretary of the Air Force announced the termination of Project BLUE BOOK, the Air Force program for the investigation of UFOS.

From 1947 to 1969, a total of 12,618 sightings were reported to Project BLUE BOOK. Of these, 701 remain "Unidentified." The project was headquartered at Wright-Patterson Air Force Base, whose personnel no longer receive, document, or investigate UFO reports.

The decision to discontinue UFO investigations was based on an evaluation of a report prepared by the University of Colorado entitled, "Scientific Study of Unidentified Flying Objects;" a review of the University of Colorado's report by the National Academy of Sciences; past UFO studies and Air Force experience investigating UFO reports during the 40s, '50s, and '60s.

As a result of these investigations and studies and experience gained from investigating UFO reports since 1948, the conclusions of Project BLUE BOOK are:(1) no UFO reported, investigated, and evaluated by the Air Force has ever given any indication of threat to our national security;(2) there has been no evidence submitted to or discovered by the Air Force that sightings categorized as "unidentified" represent technological developments or principles beyond the range of present-day scientific knowledge; and(3) there has been no evidence indicating that sightings categorized as "unidentified" are extraterrestrial vehicles.

With the termination of Project BLUE BOOK, the Air Force regulations establishing and controlling the program for investigating and analyzing UFOs were rescinded. Documentation regarding the former BLUE BOOK investigation has been permanently transferred to the Military Reference Branch, National Archives and Records Administration, Washington, DC 20408, and is available for public review and analysis.

Since Project BLUE BOOK was closed, nothing has happened to indicate that the Air Force ought to resume investigating UFOS. Because of the considerable cost to the Air Force in the past and the tight funding of Air Force needs today, there is no likelihood the Air Force will become involved with the UFO investigation again.

There are several universities and professional, scientific organizations, such as the American Association for the Advancement of Science, which has considered UFO phenomena during periodic meetings and seminars. In addition, a list of private organizations interested in aerial phenomena may be found in Gayle's Encyclopedia of Associations (edition 8, vol-. 1, pp. 432-433). Such timely review of the situation by private groups ensures that sound evidence will not be overlooked by the scientific community.

A person calling the base to report a UFO is advised to contact a private or professional organization (as mentioned above) or to contact a local law enforcement agency if the caller feels his or her public safety is endangered.

Periodically, it is erroneously stated that the remains of extraterrestrial visitors are or have been stored at Wright-Patterson AFB. There are not now nor ever been any extraterrestrial visitors or equipment on Wright-Patterson Air Force Base.

Chapter 3: UFO Events 1940s - 1950s

4. File photo of a UFO released by the U.S. govt recently.

The emergence of the Cold War confrontation in 1947 between the United States and the Soviet Union saw the first wave and evolution of UFO sightings in the modern world. The first case of a "flying saucer" in the U.S. was reported on 24th June 1947.

Gen. Nathan Twining, who was head of the Air Technical Service Command at the time, established Project SIGN in 1948, which was initially, Project SAUCER. This project's function was to collect information and distribute it to the government on UFO-linked data on suspicion that these sightings and speculations could be real.

Despite being fearful at the beginning that the objects might be Soviet secret weapons, the Air Force quickly reported in the findings that UFOs were real but easily explained that they were not extraordinary.

The report established that almost all sightings culminated from one or more of three causes which included, mass hysteria and hallucination, hoax, or misinterpretation of known objects.

Despite recommendations, the Airforce proceeded with the military intelligence control over the investigation of all sightings. However, the report did not object to the possibility of extraterrestrial existence. Following the more sightings that were recorded and identified, the Air Force kept on collecting and evaluating UFO data in the late 1940s under a new project, Grudge.

Grudge tried to lessen public anxiety over UFOs through public relations campaigns made to convince the public that UFOs had nothing unusual or extraordinary. These sightings were later explained as balloons, conventional aircraft, planets, meteors, optical illusions, solar reflections, or even "large hailstones."

However, Grudge representatives found no proof in UFO sightings of advanced foreign weapons design or development. On this, they reported that UFOs were not a threat to U.S. security. In their report, they, in turn, recommended that the project be reduced in scope and argued that the existence of Air Force official interest encouraged people to believe in UFOs and contributed to a "war hysteria" atmosphere.

During the Cold War tensions and continued UFO sightings in 1952, a new project was initiated by Maj. Gen. Charles P. Cabell, who was then USAF Director of Intelligence. Project BLUE BOOK became Air Force's main task in a bid to study the UFO phenomenon throughout the 1950s and 1960s. The identifying and explaining of UFOs fell on the Air Material Command at Wright-Patterson. With a small staff, the Air Technical Intelligence Center (ATIC) tried to persuade the public that UFOs were not extraordinary.

MJ-12 Report: Majestic 12

Just as the Blue Book, many of the inquiries have been sent to the National Archives for documentation and information about "Project MJ-12." These inquiries concern a memorandum from Robert Cutler to Gen. Nathan Twining, dated 14th July 1954.

This particular document contains problems for reasons that include:

The document is located in Record Group 341, entry 267, whose series is supposedly filed by a Top Secret register number; however, this document does not bear such a number.

While there are no other documents in the folder regarding "NSC/MJ-12", the document is filed in the folder T4-1846.

The records of the Secretary of Defense have been searched by the staff of the National Archives, and no further information has been found on this subject. Other searched areas include the Joint Chiefs of Staff, Headquarters U.S. Air Force, and other related files.

No information, however, was and has ever been produced by the U.S. Air Force, the National Security Council, or the joint Chiefs of Staff. The Freedom of Information Office (FIO) of the National Security Council literate the National Archives that "Top Secret Restricted Information is a theory that never happened nor looked into at the National Security Council until the Nixon Administration.'

This particular theory and evolution of events were not as well captured during the Eisenhower Administration as the Eisenhower Presidential Library also confirmed. The pending document does not bear an official government letterhead or watermark. The National Archives and Research Administration (NARA), conservation specialists scrutinized the paper and determined it was a ribbon copy prepared on "diction onionskin."

The Eisenhower Library has evaluated a copy sample of the documents it has on Cutler's papers collection. All documents created by Mr. Cutler while serving in the NSC staff have an eagle watermark. The onionskin carbon copies would either have or completely miss the eagle watermark.

Most documents from NSC had been prepared on White House letterhead paper. The short time that Mr. Cutler had left the National Security Council, the carbon copies he had presented were prepared on prestige onionskin. There are no records of an NSC meeting that was held on 16th July 1954. None was ever found by the National Archives after searching the official meeting minutes files of the National Security Council.

There was no record of MJ-12 nor Majestic that was ever found after a further search of all National Security Council Meeting Minutes for July 1954 were traced. There was No listing for MJ-12, Majestic, Unidentified Flying Objects, UFO, Flying Saucers, or flying discs after the social branch searched the indices of NSC records.

Native American Journalists Association (NAJA) found a memo in a folder titled "Special Meeting 16th July 1956," which indicated that the National Security Council was to be called to a defense exercise on 16th July 1956.

From a letter addressed to the military from The Eisenhower Library, that had the dates; 16th July 1957:

"President Eisenhower's Appointment Books contain no entry for a special meeting on 16th July 1954, which might have included a briefing on MJ-12. Even when the President had 'off the record meetings, the Appointment Books contain entries indicating the time of the meeting and the participants...

The Declassification office of the National Security Council has informed us that it has no record of any declassification action having been taken on this memorandum or any other documents on this alleged project ..."

Robert Cutler, at the direction of President Eisenhower, was visiting overseas military installations on the day he supposedly issued this memorandum--- 14th July 1954. The Administration Series in Eisenhower's Papers as President contains Cutler's memorandum and report to the President upon his return from the trip. The memorandum is dated 20th July 1954 and refers to Cutler's visits to installations in Europe and North Africa between July 3 and 15. Also, within the NSC Staff Papers is a memorandum dated 3rd July 1954, from Cutler to his two subordinates, James S. Ia and J. Patrick Cone, explaining how they should handle NSC administrative matters during his absence; one would assume that if the memorandum to Twining were genuine, Lay or Cone would have signed it."

The National Archives has failed to trace documentation among the Project Blue Book records, which discuss the 1947 incident in Roswell, New Mexico.

Sheila Widnall, the then Secretary of the Air Force, announced on 8th September 1994 that the U.S. had completed its study of tracing records that related to the alleged 1947 UFO incident near Roswell.

The supporters and researchers of UFO claim that an extraterrestrial spacecraft and its alien occupants were retrieved near Roswell, New Mexico, in July of 1947 and that this fact was kept from the public.

The General Accounting Office (GAO) conducted an audit in February of 1994, At the request of Congressman Steven H. Schiff (R-NM), to trace all records relating to the "Roswell Incident" and to substantiate if these records were properly handled. When the GAO audit was completed, the results were issued by the Headquarters, U.S. Air Force, in 1995. The publication is titled "The Roswell Report: Fact vs. Fiction in the New Mexico Desert." The GAO audit involved several government agencies. However, it chose to focus on the Airforce.

The Air Force initiated a systemized search of current Air Force offices as well as several archives and record centers which might help explain the incident in a bid to support the audit.

Air Force officials also interviewed several persons who may have known about the events. Before, people with any previous scrutiny were released of their obligations by Secretary Widnall to create room for their statements to be considered. The Air Force research failed to trace or develop any information that the "Roswell Incident" was a UFO event and that there was any manifestation of a "cover-up by the government."

According to the information obtained through exhaustive records, the materials recovered near Roswell were from a balloon device that had been used in a classified project. There were no records that showed or pointed out that there was the recovery of alien bodies or extraterrestrial materials. There are no records that may have suggested or hinted that the recovery of alien bodies or extraterrestrial materials had been found.

All documentation related to this case is now declassified, and the information is released for public access. More documentation has also been turned to the Air Force historian. The subcommittees chaired by Moss sent out thousands of requests that had been submitted from the public, the newspapers, and members of the Congress on topics ranging from Amelia, Ballistic missiles to frozen foods.

Two organizations of the many that contacted the FOGI Subcommittee stand out: Flying Saucers International and the National Investigations Committee on Aerial Phenomena.

Years before, the U.S. and the Soviet Union spent a lot of resources to develop programs that would send defense technology and eventually people into outer space. By mid-century, whatever existed beyond Earth's atmosphere seemed within reach, and the idea that UFOs zoomed around the galaxy got the attention of the public.

Most people who believed in UFO phenomena also got convinced that the Air Force knew about them, and that the military had resolved to keep them a secret. Anxious Americans considered this a major problem and questioned whether the Russians had already accessed the extraterrestrial technology and were using it against the United States. They further questioned whether the defense personnel needed confirmation of Unidentified Flying Objects, with the training which would enable them to differentiate the UFOs from planes and missiles in a bid to avoid the accidental war with the Soviet Union.

Many of these requests were about a specific report on UFOs that the Air Force created and called "Project Blue Book, Report No. 14."

The report had been written to substantiate whether 'flying saucers' represented technological developments not known to this (United States) country," and whether they provided explanations for why purported UFOs were unknown.

In 1955, the Air Force declassified the report, but many felt the cover-up had done a lot. In one instance, a citizen lamented that, "the government at Washington has evidence of the arrival of space travelers to the Earth, and it is serving no good purpose to refuse to confirm their arrival. The government's confirmation would clear man's fears concerning them and allow them to open their minds and hearts and welcome them so that we may profit by the new ideas they are bringing."

As one of the hundreds of subjects, Moss and his committees probed in all bureaus and all levels. This, however, bore no results from the committees that, to date, did not confirm or refute the existence of UFOs. The purpose of these committees was, however, to ensure that the public had the most information available to them.

Roswell New Mexico

In July 1947, something happened northwest of Roswell during a severe thunderstorm.

Events that surrounded the crash and recovery of a U.S. Army Air Forces high-altitude balloon had scaled to a conspiracy theory that involved UFOs. The U.S. military stimulated the interest by first claiming that the debris that was recovered was from a "flying disc." Later, they announced that the wreckage belonged to a weather balloon. In 1994 it was revealed that the balloon was part of the top-secret project, which sought to detect Soviet nuclear bomb tests.

However, that revelation did little to end the conspiracies. In the same year, the U.S. and other countries were in the middle of a "flying saucer" obsession when people reported seeing strange objects in the sky that they claimed were spacecraft piloted by aliens.

The material discovered by W.W. Brazel included tinfoil, rubber strips, and sticks. After collecting the wreckage, the Roswell Army Airfield gave a press release that admitted the recovery of a flying disc from a local ranch.

1940: Foo Fighters

While World War 2 was nearing the end, mission updates from the 415h Night Fighter Squadron took a mysterious turn.

One night of November in 1944, a Bristol Beaufighter crew, pilot Edward Schlueter, Donald J. Meiers, a radar observer, and intelligence officer Fred Ringwald were flying along the Rhine north of Strasbourg. The trio described seeing "eight to 10 bright orange lights off the left-wing, that flew in the air at high speeds." These incidents were, however, not captured by airborne radar or ground control.

"Schlueter turned toward the lights, and they disappeared, later they appeared farther away. The display continued for several minutes and then disappeared." the report read. Meiers gave these objects a name: "foo fighters."

Reports kept flowing in describing the objects that flew alongside aircraft at 200 mph as red, or orange, or green and that they appeared singly or with as many as ten others information, and they often outmaneuvered the airplanes they were chasing. They never showed up on the radar.

Explanations have, however, been offered by military aviation techs, conspiracy theorists, and psychologists, but none was found to be credible by the airmen, and they never believed that they had been hallucinating due to weariness.

Since the lights did not cause any damage, the pilots were not sure that they emanated from a remote-controlled German secret weaponry. A phenomenon known as St. Elmo's fire, when discharge of light from sharp objects in electrical fields creates visual anomalies was deemed to be unlikely since the foo fighters exhibited such extreme maneuverability. Officers were sent to investigate by the Army Air Command. Unfortunately, the research was lost after the war ended.

Six top scientists were convened in a panel by the CIA in 1953. They were experienced with experimental aviation technology and were called in to determine if the lights may have involved a national security threat. The Robertson Panel offered no official conclusion.

"I think the foo fighters didn't show up on the radar because they were plain light," Howard said. "Radar had to have a solid object. If there were any bogey out there, the pilots would be able to tell." Ziebart, the historian, did not explain either.

Ghost Rockets – Scandinavia

In 1946, other UFO sightings were witnessed, this time in Europe over the Scandinavian countries where they were called Ghost Rockets. The sightings did not start in the U.S. with the 1947 sighting by Kenneth Arnold.

A year earlier, waves of UFOs were seen in Sweden, Norway, Denmark, and Finland. By this time, they had not been referred to as Flying Saucers or UFOs. These terms came into existence in 1947 and the latter in 1952. The Swedish government held back information and released them 40 years later.

9th June 1949

A sharp, clear light dashed over Helsinki, Finland. It was accompanied by a smoke chain and the sound of thunder. The luminous chain lasted for ten minutes, and the same incident repeated the next night. During the second night, the Ghost Rocket turned and went back in the direction from which it came, hence ruling out the possibility of it being a natural phenomenon like a meteor or asteroid.

On 12th June 1946, the Swedish Defense Staff directed the military personnel to report their sightings through official channels and admitted to knowing of the phenomena since May. On 9th July 1946, more than 200 Ghost Rocket sightings were reported, most being described as tubular or spindle-shaped that flew low and at low speeds and produced little or no sound at all.

After the incidents, the Swedish government formed a special committee to investigate. A week later, James Forrestal- American Secretary of the Navy, traveled to Stockholm to meet with the Swedish Secretary of War. This followed with the reporting of over 300 cases of strange sightings in the Stockholm area on 11th August 1946. UFO researchers later found more than 1,500 reports that had been secretly collected and finally opened by the Swedish government in 1984 as 'Ghost Rocket Files.'

Washington DC UFO incident

5. Air Staff / National Archives, Records of Headquarters U.S. Air Force

The 1952 Washington D.C. UFO incident is also referred to as the "Washington flap" and the "Invasion of Washington". This involved a procession of UFO incidents that had been reported from 12th July to 29th July of 1952 in Washington DC.

The most reported sightings occurred on consecutive weekends of July 19–20. Then after that on July 26 and 27. UFO historians developed names and theories on this, but according to Curtis Peebles, he termed and referred to this incident as "the climax of the 1952 (UFO) flap." On 19th July 1952, on a Saturday, the crew at Washington National Airport came across an unusual sighting, which ran in the headline of the following day's news reports.

"Saucers Swarm over Capitol," the front page of one newsroom, the Cedar Rapids Gazette, read. That headline and many others on the day it read created anxiety as to why UFOs groupings would spend their weekend hovering the D.C., five years down after the famous Roswell, New Mexico incident. While other sightings may have eyewitness testimonies and pieces of evidence that eyewitnesses produced, the D.C. one was captured on the radar by the National Airport, and they were undeniably present.

"There was an attempted intercept, but the planes got there, and everything was gone." Reports from the Air Command center read. A week later, on 26th July, which was still a Saturday, the crew was working at the radar facility; they were prepared for the possibility of seeing UFOs. Luckily, the blips reappeared on the radar screen, and the crew called for another interception.

The incidents over D.C. were one of many UFO sightings seen throughout the country that the Air Force had already investigated via Project Blue Book. In the end, Project Blue Book had three major conclusions:

1. None of the UFO sightings was a threat to United States national security.
2. The technology presented in all UFO sightings did not suggest otherworldly evolutions to aircraft tech of the days.
3. No evidence suggested that UFOs were from outer space.

Fiorentina Stadium UFO incident

On 27th October 1954, a supernatural episode occurred in Tuscan before 10,000 people who watched Fiorentina play rivals Pistoiese in a friendly when a disturbance in the sky captured everyone's attention.

An unidentified flying object appeared, and the thousands in attendance were astonished by the disc-shaped object clouds that were in the skies above them. This lasted almost 15 minutes, and the egg-shaped spacecraft zoomed off just as fast as it had appeared, but not before emitting silver flakes of a stringy-type substance, which covered the landscape, including the trees.

In the days before the match, other sightings in the region had been spotted but didn't deter experts from thinking those in attendance might have had too much Chianti Classico that afternoon.

"The whole UFO phenomenon is nothing but a myth, magic and superstition wrapped up in this idea that somehow aliens are either coming to save us or destroy us; when I first looked at this case, I originally thought it was a fireball ... but it became apparent that this was caused by young spiders spinning very thin webs. " Astronomer James McGaha said during a BBC interview. "

Project Blue Book

The Air Force investigated UFOs under Project Blue Book from 1947 to 1969. The project was later ceased on 17th December 1969. In total, 12,618 sightings were reported to Project Blue Book, with 701 cases remaining unknown.

The decision to cease UFO investigations was based on an evaluation of a report prepared by the University of Colorado; "Scientific Study of Unidentified Flying Objects," which was a review of the University's report by the National Academy of Sciences, previous UFO studies, and Air Force experiences while investigating UFO reports.

These investigations, which included experiences and studies since 1948, concluded that:

1. No UFO reported, investigated, and evaluated was ever an indication of a national security threat.
2. No evidence discovered by the Air Force suggested that sightings categorized as "unidentified" were technological developments. None either suggested that they were beyond the range of modern scientific knowledge.
3. No evidence indicated that sightings categorized as "unidentified" were extraterrestrial vehicles.

After its termination, the Air Force regulation establishing and controlling the program for investigating and analyzing UFOs was canceled. The documentations were then permanently relocated to the Modern Military Branch, National Archives and Records Service. Nothing has since occurred since the termination of Project Blue Book to support the resumption of UFO investigations by the Air Force. With the decreasing defense budgets, it is unlikely for Project Blue Book to be resumed by the Air Force any time soon.

Chapter 4: UFO Events of the 70s, 80s, and 90s

Tehran, Iran UFO Incident

In September 1976, two Iranian Air Force F-4 Phantom II fighter jets flew out to investigate and intercept UFO over the capital of Tehran. As they approached the UFO, it is reported that the planes suffered various equipment malfunctions. Avionic systems such as navigation, weapons, communications, and more failed on approach, which seemingly fixed themselves upon departure from the object.

The Iranian Air Force had earlier received several reports of an object in the sky that was brighter and larger than a star, and commanders ordered that the object be commandeered. The first jet got within 25 miles radius of the said object and started experiencing odd electronic failures. It then had to change course and abandon the target. A while later, the second jet approached and was able to get radar at a 27-mile range. The object's radar signature was a 707 jetliner, but with its brilliant and many smaller colored lights, it was immediately resolved as not to be a jetliner. The crew of the second F-4 discerned how a smaller object detached itself from the larger one and chased the jets.

The second F-4 experienced electronic failures when it wanted to engage the detached smaller objects with its radar-guided missiles and was forced to turn back. At this time, a second smaller object detached itself from the main UFO and fell to Earth. The area where the second ejected object landed later was identified as a dry bed lake. There was, however, no trace of it on the ground. It was then reported that the objects that most people came to see were a trail of meteor showers which often occur annually.

On equipment malfunctioning, the commission responsible reported that only one plane had a technical equipment malfunction. The fighter pilots and their crews, as well as the Iranian defense authorities, got persuaded that the Unidentified were extraterrestrials. They went ahead and referred to them as fireballs. The possibility of an outstanding, fast Soviet jet was taken into account, but Soviet jets do not glow brighter than stars.

The pilots described the main UFO as saucer-shaped with a dome on its top and that it appeared tubular from certain angles. However, they could not determine the sizes of the objects since they produced sharp blinding lights but agreed that they traveled as fast as 2000 to 3000 mph.

Author Brian Dunning said of the incident, "Once we look at all the story's elements without the presumption of an alien spaceship, the only thing unusual about the Tehran 1976 UFO case is that planes were chasing celestial objects and had equipment failures. There have been many cases where planes had equipment failures, and there have been many cases where planes misidentified celestial objects. Once in a while, both will happen on the same flight."

In the third scenario, the control tower equipment failed when the third object flew over the base control. The same way they failed was similar to how the planes, which flew a couple of miles from the objects, failed. The U.S. Air Force was interested in the occurrence. The U.S. Defense Intelligence Agency cataloged the incident and sent reports to the White House, the military, the CIA, and the NSA. Unfortunately, after the Iranian Revolution occurred in 1979, all access to Iranian records was closed off.

The Cerenchovo (Siberian Russia) UFO Incident

On 27th September of 1989, according to the official reports, tall aliens with small heads and three eyes showed up in the city of Voronezh, arriving, accompanied by robots, in a shiny ball aircraft.

The following is an excerpt from an Associated Press translation of the Russian report:

"Scientists have confirmed that an unidentified flying object recently landed in a park in the Russian city of Voronezh. They have also identified the landing site and found traces of aliens who made a short promenade about the park." They left behind them 'two pieces of unidentified rocks,' made of a substance that 'cannot be found on Earth'."

The New York Times reported that that was not all: on 23rd October of 1989, they published the following:

Earlier in the year, the newspaper Socialist Industry reported an "encounter" between a milkmaid in the region of Perm and a cosmic creature that looked like a man but was "taller than average with shorter legs."

Last week the Soviet newspaper Komsomolskaya Pravda *declared that not only had an Abominable Snowman been caught stealing apples in the Saratov region, but researchers had "registered the influence of energies" at a site in Perm, leading a geologist to conclude that they had discovered a landing field for flying saucers.*

The same story transcribed a telepathic discourse between Pavel Mukhortov, a journalist from Riga, and an all-too-knowing extraterrestrial:

"Where are you from?" asked Mukhortov.

"The Red Star of the Constellation of Libra is our home."

"Could you shift me to your planet?"

"That will mean no return for you and danger for us."

"What danger?"

"Thought bacteria."

To the chagrin of Soviet scientists, the thought bacteria are everywhere."

Writer Howard G. Chua-Eoan stated that there was a reason for TASS and other Soviet news outlets going insane. The policy of openness in the media was new. Publishers were testing the waters with how far they could go, especially at a time when hope for the Soviet Union was waning. Stories and theories of aliens and transcendental creatures provided something a little less depressing to think about.

Though several educated Soviets strongly denied scientifically unproven trends in the state media, reports of UFOs were not the only fake documents for them to be dealing with. "They've been feeding us rubbish about the dream of Communism for years, and we now see they were lying," a Soviet source told Time Magazine in 1989.

The Russians have strange histories with the Unidentified Flying Objects. Some Ufologists believe that nearly all of the UFOs encountered by Russia and the Soviet Union over the years were probably American spy planes.

They, however, insist that one of them landed, and its crew decided to step out and stretch their legs. The New York Times quoted police Lt. Sergei A. Matveyev in the late 1980s, who had sworn to have seen the spaceship with three-legged creatures, saying they landed in a park in Voronezh, a Russian city, in the late September of 1989. The people living in Voronezh reported a deep red ball, around 10 feet in diameter, landing in a park 300 miles from the capital of Russia.

"It was not an optical illusion," he told the Russian TASS News Agency. "It was certainly a body flying in the sky. I thought I must be really tired, but I rubbed my eyes, and it didn't go away."

It is reported that after the disk landed, it opened, and three-eyed creatures that stood nine feet tall, dressed in silver overalls and bronze boots came out. The creature was in the company of another and a robot.

They then took a triangle formation with the robot in the center, and it came to life. A boy who witnessed this began to scream in terror. That paralyzed him. The aliens briefly disappeared and came back with a gun-like tools and shot the boy. This made the boy disappear and reappear after the spacecraft had left.

The residents of Voronezh reported more sightings of the alien ship between Sept. 23-27. This was, however, not the case for Soviet investigators. On visiting the scene, they could only read the radioactive date of 23. The children who witnessed the ship's landing drew a banana-shaped object that left an "X" sign in the sky, although they had been separated into different rooms.

As for the abducted boy, he remembered nothing about the incident. The local interior minister said they would dispatch the Red Army had the aircraft been sighted again and promised that they would meet the force of the Soviet Union and the United States, should they have invaded again.

Bob Lazar and Area 51

Area 51 came to be famous when Bob Lazar, a former government physicist, went public in 1989 with his experience of reverse-engineering an alien ship for the U.S. military. This was due to happen at a top-secret Air Force test facility across Nevada. For three decades, Lazar was largely unknown until filmmaker Jeremy Kenyon Lockyer Corbell found him to re-explore Lazar's story, as well as explain why Area 51 still attracts a lot of public fascination.

He has been noted as making some of the most unflinching claims in recent history, the famous being the claims that put area 51 in the limelight as a test facility for UFOs, aliens, and other top-secret, government-science experiments. Although UFOs and alien life were popular before Lazar came to prominence, his accounts have kept theorists busy for decades.

In 1989, Bob Lazar revealed shocking details about his previous employer. Lazar worked with the Air Force from 1980, in the south of Area 51 called S4. During his tenure, Lazar was privy to the amazing technology that the military and the government had been working on. Lazars's role at the time involved reverse-engineering extraterrestrial space craft. He was handed files and documents on the craft that he was handling. This specific craft was allegedly a part of an archeological break-up. At the base, scientists were grouped depending on their expertise and the aircraft they were working on.

As for Lazar, he belonged to the propulsion group while his colleagues and researchers studied the flight controls, navigation, lift, and other technologies they had been assigned. Bob Lazar described his experience at S4 on the craft he had been assigned to work on and a second craft which he came to contact with at the base, a live alien, and more interesting assertations.

The Public Revelation

According to Lazar, he discovered that the Air Force was scheduled to experiment with the re-engineered craft in the middle of the night at Groom Lake. He called out his friends to come out to the desert to witness the trials. He was, however, discharged from the base after his employer learned of Lazar's U.F.O. "show guests." Shortly after his dismissal, Lazar was a guest to C.B.S. reporter George Knapp in Las Vegas. For the first time since his dismissal, Lazar provided details on his experience at area 51.

Since his revelations, Lazar became inactive for some time until 2018, when 'Bob Lazar: Area 51 and Flying Saucers' director Jeremy Corbell put him back in the public eye. This documentary landed him more attention along with when he appeared as a guest on the Joe Rogan Podcast. This interview has been viewed and listened to by millions of people, an even larger audience than his interview with George Knapp.

Lazar's Claims and Evidence

In trying to justify Lazar's claims, a few things stood out; before he went public with the information on area 51, the public had no information about the secret testing that occurred at the base. However, from his claims, area 51 gained attention which it carries to date. Strangely, it has been revealed that Area 51 has been a testing base for government secret projects, including the Stealth Bomber aircraft. While at the base, Lazar took groups of people to experience the tests that were privy to only a few people.

The second oddity; Lazar claims that he gained access to the base through unique security access systems that include hand sensors that analyzed the bone density to analyze and identify an individual. In his defense, the Air Force acknowledged that such scanners were used at the time of Lazar's employment. However, public knowledge of such technology at the time of Knapp's interview did not exist.

Element 115

One fascinating proof that supports Lazar's claims and that makes his argument justifiable include Element 115, which according to Lazar, involves a main piece of the propulsion system from the craft which he was working on, called 'element 115'. In the periodic list of tables, element 115 is referred to as 'Moscovium.' In 2003, the Russian scientists synthesized element 115 in a lab and named it Moscovium.

Until 2003, which was 14 years since Lazar claimed to have been researching the element Moscovium for the U.S. military, it had never been heard of.

Lazar's Story and Shortcomings

As much as there is compelling evidence that supports Lazar's experiences and suggests that he was telling the truth about Area 51, there are equally supporting theories that suggest he fabricated his own theory.

To start with, Lazar's arguments have been changing over time. There are a few details that changed in the second interview with Rogan that he had not mentioned at Knapp's. Initially, Lazar said he had seen a live alien being interrogated by a number of agents but later retracted his statement and said it was a mannequin that was made and looked like it would have been alien.

Secondly, he claimed that his educational and professional backgrounds were impressive; however, his records on previous employers and schools did not exist, except the Air Force, which accepts that Lazar was once in their payroll, and in fact, includes an employee phonebook that has Lazar's records and information contacts at the time.

One thing for sure, Lazar was unmistakable. He did not offer any vague and hypnotic hints about the U.S. scientists may have "foreign materials" that require advanced testing and examination of whether they belong to the alien origin. He laid out clearly that scientists had nine alien detainee crafts, which they had been studying for more than thirty years. He further reiterated that he worked on these alien machines.

Area 51

There may be Tang courtesy of the space program, but the question remains; where did the idea on innovations like the Stealth Fighter and Kevlar come from? Could it be aliens? Remains from crashed U.F.O. space craft have been claimed to be stored at Area 51 by the conspiracy theorists - a base that the government uses to reverse-engineer aliens' hi-tech gear.

This information comes from different U.F.O. sightings in the highlighted area and the testimonies from a retired Army Colonel who claimed to have received access to extraterrestrial materials that had been collected from a crashed alien spacecraft in Roswell, New Mexico. Other theorists believe that the government is studying time travel, also known as Dreamland or Groom Lake at Area 51.

Advanced aircraft like Stealth Bombers and patrol planes, amongst other weapons, have been developed by the government at the nearby Nellis Air Force Base. The claims by the government that Area 51 is classified for purposes of National Security are seen as proof of the government concealing the truth about aliens and exotic space crafts.

The Secret Base

Before the Second World War, 'Area 51' was used for mining silver and lead. After the war began, the remote area was repossessed by the military, which they used to carry out their research on nuclear and weapons testing.

In 1955, the then Central Intelligence Agency Director, Richard Bissell, together with Kelly Johnson, an aircraft designer, set aside an airfield at Groom Lake as their headquarters to develop spy reconnaissance planes during the Cold War. This area was later renamed and labeled as site Area 51and added to the existing Map of the Nevada Test Site by the Atomic Energy Commission.

The U-2 plane was developed by engineers within **eight months.** These planes could soar at an altitude of 70,000 feet which was higher than any other aircraft at the time. With these developments, pilots could fly conscientiously above Soviet radar and enemy aircraft.

In 1960, a U-2 was shot down by a Soviet anti-air missile. This prompted the C.I.A. to begin developing the next generation of spy planes at Area 51. This became the 'titanium-bodied A-12.' Almost undetectable to radar, the A-12 flew across the continental United States in 70 minutes at 2,200 miles an hour. These planes came, fitted with cameras that could capture images and objects that were one foot on the ground from a height of 90,000 feet.

With more than 2,850 takeoffs by the A-12, and other high-tech flights out of Area 51, U.F.O. sightings around the area surged. In her book on Area 51, Annie Jacobsen quotes a source saying: "The aircraft's titanium body moved as fast as a bullet. It would reflect the rays from the sun in a way that could make anyone think they were Unidentified Flying Objects."

In 1989 when Bob Lazar came forward with his story, Area 51 was fully associated with aliens. His claims involved housing and studying alien space crafts, and recreating technology for military use.

It was verified that indeed engineers at Area 51 at the time studied and re-created advanced aircraft and not so much alien-like, which were achieved from other countries. However, the assumption that Area 51 houses alien spacecraft still remains.

Where is the Truth?

In 2013, the government befittingly acknowledged that Area 51 existed for the first time and the C.I.A. declassified tones of files on the evolution of the U-2 and A-12. Initially, details were scarce and hard to verify, but natives knew something odd was happening in the desert.

Area 51 remains an active base that develops unimaginable military technology. As much as details on which and which aircraft were being developed before the 1970s, and in support of the released documents by the C.I.A., historians still have no clue of any other generations since, and that remains a top-secret mystery. As much as it is not clear when the current work on mystery will be released to the public, it might certainly take a few more decades.

The site stands as a pedestal of U.S. alien mythology. The **"Storm Area 51" event,** which was inspired by a recent interview between Lazar and Joe Rogan on a popular podcast, has since deformed to a festival celebrating anything that is alien. Area 51 has drawn believers and intellectuals who visit the small but thriving trail of alien-themed museums, restaurants, motels, parades, and festivals, even without the festival.

Lazar's story remains phenomenal. Had it been true, it would certainly change history, the future, and the world we live in today. It could be a gigantic discovery in human history. However, with the nature of the story, it is difficult to judge its veracity. There is plenty of evidence to either go by Lazar's theories and claims or completely ignore the stories as put forward by him.

1994 Ruwa, Zimbabwe Incident

A group of 62 children at Ariel School in Ruwa, Zimbabwe, was playing in the field when they saw U.F.O..O. coming closer to them during their mid-morning break. The children, aged between eight and twelve, said they saw a U.F.O. and "strange beings." From what they reported, the children who witnessed the 1994 incident claimed the U.F.O. had given those warnings about the future of tech, which the modern world currently uses.

An American filmmaker, Randall Nickerson, was fascinated by this story and hence decided to investigate it.

During the morning break, while children were in the playground, something suddenly flew nearby, coming to the edge of the school playing ground. Before it landed, the children narrated incidents of how it levitated off the ground and the bushes around and the hill. Out of curiosity, they ran towards the craft to see what it was. Their version explained a creature that walked around the craft while the other was near them. They told this story that the creatures wore tight black suits with big eyes on their heads.

Standing in front of them, they were put in communication like creatures. This scared them when the bell rang, awakening them to find that they were in front of these creatures. While in class, they kept talking about what they had seen while on the playing field in the morning.

In the beginning, the teachers doubted what they were talking about, saying it was their imagination. Logically, when a group of kids tells their teachers of something they have seen, the teachers have to believe them. They are innocent children who all saw what was on the playing field that morning. The children got silent in class, and the teachers wondered what to do.

In the afternoon, the children maintained their stories and even told their parents what they had seen back in school that morning. They made their parents think they were crazy. However, they were concerned, prompting them to call the school to validate the claims of their children.

After separating them, the children were told to draw what they had witnessed, and all drew the same thing. Each drew an aircraft that resembled the letter 'X", which floated in the air and emitted a trail of smoke.

6. Drawings by pupils at Ariel School in Zimbabwe of the aliens' visit to their school.

John E. Mack, a Harvard professor of psychiatry through an invite by the B.B.C. crew, visited the school in November 1994 and conducted televised interviews on the incident. Each child was interviewed separately. However, to his surprise, he discovered the children's story was true.

Mack observed their voice tones and body language. Everything was consistent. From a statement one of the students said, he quoted that; "we should not rely on technology to try to find them." This fascinated Mack, and he wondered what a child of little age would understand about technology.

Before he died in 2000, Mack had publicly confessed his experiences with foreign beings in the past. He had spent about a decade and a half of her life researching U.F.O. sightings on the African continent. He worked and carried out his research for the Mutual U.F.O. Network in every newsletter that U.F.O. Afrinews distributed.

From an excerpt published by Andrew Owen on the interviews he collected from the witnesses in 1994, he indicates that he brought along a printout from one of the issues which he opened before Sarah, one of the witnesses, an article that had been written before.

The reading is: "U.F.O. Wave in Zimbabwe":

"14th September 1994; a Wednesday was an exciting night. At about 20:50 to 21:05 hours, a fireworks display of some Grandeur Materialized in the clear skies of this part of Southern Africa."

Soon, astronomers reported that the 'fireworks display' that had been seen in countries including Zambia and Botswana were Meteor showers. However, as for Hind, she recorded dozens of reports of rectangular-shaped fireballs with fire which were flanked by two smaller shaped fireballs.

Around the same time, she recorded reports of alien sightings citing a young boy and his mother who claimed to have seen a daylight sighting and a trucker who had witnessed strange essence on the road at night. In September, she received the Ariel School report, which she named case 96, and described it as the most interesting U.F.O. story of all ages.

Childhood Reminiscence

Hind's account closely reflects Sarah's reminiscence. Hind records that the children were out for their morning break at 10 am, who were then attracted to an area with shrubs and bushes in their playing field; the bushes high enough to hide a child should he or she choose to walk through.

The recollection has that all teachers were in the staff room having a meeting and the only adult who was out at the time was the tuck-shop mistress, who was soon flooded by children who claimed to have seen "three or four things which came into the bush are, the disc-shaped objects coming in along the power lines and finally landing among the trees. The children were scared as well as curious to know what they were.

In the reviewed incidents, the analysts focused on occurrences that involved some form of physical evidence, which included photographic, radar, vehicle interference, aircraft equipment interference, inertial effects, traces on the ground, destruction on the vegetation, effects on the witnesses, and the debris.

The main concern includes that these U.F.O. happenings could have repercussions on people's health, especially the eyewitnesses. As already reported from U.F.O. reports, some have reported radiation-type injuries.

According to scientists, some recorded incidents may have been a result of rare natural phenomena. These may have been electrical activity beyond the equivalents of thunderstorms and radar conduits. However, as from the interviews relating to the Ariel School incident, researchers and scientists have concluded that these sightings are true, face the challenge of being explained in whichever fashion.

Chapter 5: The Disclosure Project

Steven M. Greer took the podium on May 9, 2001, at the National Press Club, in Washington, D.C., in a bid of the truth about unidentified flying objects. Greer, an outspoken ufologist, knew that the government had long held information on its association with alien visitors from the American people. In 1993, he had founded the Disclosure Project in an adventure to penetrate the study of conspiracies.

On a specific day, Greer's retribution featured twenty speakers for whom, in support of his claims, he provided four-hundred-and-ninety-two-page documentation, which he had named Disclosure Project Briefing.

He then prepared another ninety-five-page summary of the Disclosure Project Brief for public officials who were too busy to absorb such a large tract of suppressed knowledge. According to Greer, untold numbers of alien crafts have been observed in our planet's airspace. While working at the base, they were able to reach extreme momentums with no physical means of lift or propulsion, as well as perform incredible maneuvers at g-forces.

Some of these exotic spaceships had been "brought down, recovered and studied since the 1940s and possibly as early as the 1930s." Efforts to reverse engineer such phenomenal machines had led to "remarkable technological breakthroughs in the generation of Energy."

Most of these performances had been classified as "cosmic top secret," a seal of approval, "thirty-eight levels" above that naturally accepted to the Commander-in-Chief. Greer wondered why such transformative technologies had been hidden for decades, but as obvious as it was, the world was ruled by the social, economic, and geopolitical order, which at his reawakening, was at risk.

Since the postwar, the idea that aliens had haunted our planet circulated amongst ufologists. When George Adamski declared to have met with a race of kindly, Nordic-looking Venerians who were bothered by the domestic and galactic effects of nuclear-bomb tests, in 1947, an alien spaceship is claimed to have crashed near Roswell, New Mexico. Conspiracy theorists believed that, in a way, humanoid bodies had been recovered there and that the crash debris had been consigned to private military contractors, who rushed to unlock alien paraphernalia before the Russians.

Upon the fall of the Soviet Union, the recovered documents suggested that the anxiety about an arms race amplified by alien tech was mutual. These had been covered up by Majestic 12, as the ufologists claimed, a surreptitious, sovereign organization converged under an executive order issued by President Harry Truman.

After planning to pitch and level out with Premier Khrushchev, President Kennedy was assassinated. He had disclosed this to Marilyn Monroe, hence sealing Monroe's fate. Representative Steven Schiff died of cancer despite spending years trying to unravel exactly what could have happened in Roswell, N.M.

The "Executive Summary" that Greer prepared was detailed; only discerning readers could find within it remedies to many of the frequently asked questions about U.F.O.s. Concerns like why these extraterrestrials were so elusive and whether we were being monitored by the aliens, and the reasons for that came forth. Greer, however, gave responses in his summary to answer his readers that some concerns as to why the aliens monitored humans was because they were discomfited by humans' desire to "weaponize the skies."

In their defense, Greer argued that these extraterrestrials are friendly and that any civilization capable of routine astral travel could put an end to our civilization in a jiffy if that was their intention. He continued that while we still breathe the free air of planet earth, that is enough explanation and testimony to the friendly nature of the extraterrestrial civilization.

Back at the conference, Greer hinted at the need to disclose the details surrounding what he then referred to as "the little green men." He cited:

"I know many in the media would like to talk about little green men... the subject is laughable, yet so serious...grown men in the congress weep, and some have asked me, 'what do we do?' this is what we should do; we have to see that this matter is properly disclosed."

Ufologists have perennial faith in the peril of disclosure, an art term for the government's ecstatic confession of its extreme U.F.O. knowledge. Years after Greer's press conference, the expected announcement on U.F.O. was suspended by the events of September 11th.

Greer then issued a "Special Presidential Briefing for the then U.S President, Barack Obama," in 2009. He claimed that the non-intervention of Obama's antecedents had "led to an illiterate crisis that will be the greatest of Obama's Presidency." To counter him, Obama's response remains unknown, but in 2011 ufologists classified two appeals with the White House. The Office of Science and Technology Policy said that there was no evidence to determine whether extraterrestrials were actually present or whether they had contacted any of the members of the human race.

Government Cover-ups

Most organizations have come forward pushing for their governments to release information on Aliens and other extraterrestrials. These include Dr. Steven Greer, who has specifically pointed out the United States government for covertly studying the extraterrestrials since the 1940s, calling for the administration to put public its relationship with aliens and other exotics from infinity.

The Disclosure Project, which was founded by Greer, has more than theory and has compiled testimonies of more than 400 government officials.

According to Disclosure Project's website (www.disclosureproject.org), Greer states to have concluded his research stating that he has established, without doubt, the truth about the extraterrestrials life, U.F.O.s, and their vehicles and evolved energy and propulsion technology which is as a result of the vehicles he has studied.

A Professor at the University of New Hampshire, Dr. Ted Loder, called the research first-rate; he has as well been involved with the project for five years. Loder has maintained that the U.S. government is withholding information about extraterrestrials and hopes that this theory would be addressed faithfully to the future generation.

During a T.V. show in one of the mainstream media, Loder stated that the students, being the next generation, should be brought to light on details of U.F.O., which have been kept under wraps for more than fifty years.

The Disclosure Project hopes that Congress will address the government's knowledge of extraterrestrial life in open hearings if it brings enough attention to the subject.

The organization, which was founded in 1993, has held claims that government agencies have intentionally deceived and denied access to the former U.S presidents and that they are aware of extraterrestrials and hence operating without the guidance of Congress. The Disclosure Project holds that "the unconditional solutions to the world's energy, pollution, and poverty problems exist within distinguished projects that need to plan divulgence and relevant codification."

Greer has assembled more than 120 hours of testimony which he has acquired through interviews with government officials. This data has been compressed to 18 hours. A significant testimony is one he had with the Army National Guard Reserves Brigadier, General Stephen Lovekin, and Gordon Cooper, an astronaut.

A natural food businessman who worked for the organization was quoted saying: "Greer knows that the world is headed for a disaster, and he believes he can do something to prevent that." Murray, 52, said the presumption of extraterrestrial life dawned on him over 30 years before.

"I do not just believe in U.F.O.s. It's true to me. Human beings are herd animals, and authorities are trying to make people believe the things that are not true. The world will completely change once the truth is revealed." Murray said.

"If there's nothing out there, then let's talk about it," Bernardi said. "It would be arrogant to think we're alone." Said Robin Bernardi, owner of a small local publishing company who has been working to push the awareness of U.F.O. existence.

Chapter 6: Recent U.F.O. Sightings

Nimitz Carrier Strike Group-2004

As one of the 21st century's most baffling U.F.O. sighting, the incidents that surround the fighter pilots and radar operators from the U.S.S, Nimitz remain phenomenal. It all started as a routine naval training drill. Among the witnesses were a highly trained military workforce, experienced radar operators, and fighter pilots who were working at the controls at the time of the sighting, or conceivably the most evolved flight technology that had ever been created.

It was on November 14, 2004, at the Pacific Ocean, about 100 miles southwest of San Diego, California. The U.S.S. Nimitz Carrier Strike Group was conducting a string of drills before they were deployed to the

Persian Gulf. In their possession, they had nuclear-powered carriers and missile cruisers U.S.S. Princeton.

At about 2 pm, two Super Hornet fighter jets from the Nimitz received an abnormal order from an operations officer in Princeton. They were airborne at the time when the pilots were ordered to stop their training operations and proceed to new points for a "real-world" mission. More fateful, the officer on the line asked if they were carrying live weapons, to which they responded that they had none.

A Puzzling Presence at 80,000 Feet

Princeton's highly state-of-the-art radar was picking up mysterious objects for several days at the time. The Navy referred to these objects as "anomalous aerial vehicles," abbreviated as AAVs. This was the term the military approved for unidentified flying objects, which had been decayed by its relationship with flying saucers, little green men, and innumerous crackpots.

According to Princeton's senior radar operator at the time, Kevin Day, his screen reflected over 100 AAVs over the week. "If you watched them on display, you mistake them for the snow falling from the sky," Day said in his recorded interview with HISTORY for the program "Unidentified; Inside U.F.O.s Inquiry."

According to the air commander of the day, the AAVs came at an altitude greater than 80,000 feet. This was far higher than commercial or military jets could regularly fly. In the beginning, Princeton's radar team did not believe what they saw, striking up the anomalies to an equipment malfunction. Nevertheless, they established that everything was working as they were supposed to and hence, began observing instances in which the AAVs fell with confounding speed to lower, busier airspace. At this point, day contacted Princeton's commander for reaction.

"I was gnawing at the bit," he says. "I really wanted to head off past the things ahead of me." Two fighters were redirected to deflect one of the bizarre phenomena. At first, when they arrived on the scene, the pilots didn't trace any flying objects. However, they recognized what Commander David Fravor, who was the lead pilot, came to refer to as Ocean "disturbance."

The water was swirling with white-like waves breaking over what seemed to be a large object just under the water. Fravor, who was the commander of the elite Black Aces squadron and a Top Gun program graduate with 16 years of flying experience, noticed one of the objects flying about 50 feet above the water. He described it as about 40 feet long, shaped like a Tic Tac candy, and with no obvious means of thrust:

"It's white with no wings. It has no rotors... what is that?"

It had swift and erratic movements that were obviously odd. Fravor described this sighting as something he had never come in contact with before: "This thing moved one side to the other, almost the same as throwing a ping-pong ball to a barrier."

The other pilot who served as a Wingman to Fravor, whose identity has never been disclosed, told HISTORY a similar account to Fravor's. At the time, the current high-ranking Navy officer was a rookie pilot. That was back in 2004.

She remembered how terrified she was as she watched how the experienced pilot tried to intercept the peculiar craft: "It was so unforeseeable, high ground, rapid pace, and momentum. I wondered; how could I possibly fight with that?'

According to Fravor, the craft ascended and came right at his plane whenever he flew around it: "suddenly, it turned and quickly accelerated, faster than anything I have ever seen. The next moment it's gone...you can't see it."

Princeton's radar was tracking as Navy jets began launching off the carrier to try and intercept the other mysterious objects. All this while, the Tic Tac accelerated into the distance.

The picture of the Tic Tac was captured by a highly sensitive infrared camera by one of the pilots who had landed and was able to track it as soon as Fravor took off. From his altitude, Fravor could not capture the encounter on video.

Nimitz Incident Prolonged Attention

The Nimitz incident seems to be known in naval circles and to some as UFO/AAV buffs. It, however, didn't come to extended public attention until 2017, after an article on the sighting was ran by The New York Times, and a video of the Tic Tac that had been shot by the Nimitz pilot was released.

The existence of a Defense Department initiative called the Advanced Aerospace Threat Identification Program, or AATIP, which was little-known about, was also disclosed by The Times. The "shadowy" operation, as the Times called it, had a budget of just $22 million, which was less than 0.004% of the Department's total budget. This was rumored to be Harry Reid, the former Senate Majority Leader's project.

Despite claims by the government that AATIP had been shut down in 2012, Luis Elizondo, who was the former director, maintained that it was still functioning. In 2017, Elizondo left the program claiming that its work wasn't being treated seriously enough within the Defense Department as it should have been.

Since Elizondo joined 'To the Stars Academy,' an organization co-founded by Tom DeLonge, who is known as the guitarist with the band Blink-182, 'To the Stars Academy' is well known as a promoter of U.A.P. investigation and exploration.

Apparently, 'To the Stars Academy' have plenty to work with. According to a former U.S. intelligence official who is now a national security adviser to the organization, Christopher Mellon, the Nimitz incident was not an inaccessible event. Incidents off the East Coast have been reported, and the 2015 case is the most recent one.

The Navy appears to be taking the matter seriously with the modern situations, despite abandoning the Nimitz case in 2004 with unyielding information. In April 2019, the Navy announced it was coming up with new guidelines that would facilitate easy reporting of any sightings on "unidentified aircraft." This move was intended to exonerate such reports and make it easier for service members to fearlessly come forward on the matter.

There exist a number of findings of unauthorized and/or unidentified aircraft entering various military-controlled ranges and classified airspace in recent years." According to a Navy Spokesperson, each and every report concerning these claims was being investigated by the Navy.

As with all things U.F.O., AAV and U.A.P., the Nimitz incident has its doubters. Some have suggested the crafts were advanced reconnaissance drones and that the churning water was caused by a submarine. But whatever the now-famous Tic Tac actually was, it's hard to dispute that the pilots, the radar operators, and the infrared camera had seen something. And chances are, it wasn't just a big breath mint.

O'Hare International Airport Sighting

7. SIGHTING OF DISC-SHAPED OBJECT OVER O'HARE AIRPORT, CHICAGO, ILLINOIS, AT 16:30 HRS. (CENTRAL) ON TUESDAY, NOVEMBER 07, 2006

O'Hare Airport is famously known by ufologists as the home of one of the most reported sightings of all times. In November 2006, a "flying saucer-like object" was spotted over corridor C of the United terminal by pilots, mechanics, and the airline management. The incident was reported by a Chicago Tribune columnist a few weeks later.

The Chicago Tribune was keen to note that the object was first seen by a ramp worker around 4:30 pm. After that, a diversity of witnesses reported that the object was dark gray in color and was 6 to 24 feet in diameter. Some said it materialized as a spinning Frisbee, while others claimed it wasn't rotating at all.

Despite the contrast, all concurred that the object was silent and emerged under the 1,900-foot cloud deck. It then scooted into the clouds and left a ring-shaped hole in the clouds as it moved. "I know that is what I saw and what a lot of other people saw as it stood out very clearly. Definitely, it was not an aircraft from the earth," A mechanic at the airport said. A manager on corridor B ran from his office after hearing the report on the sighting on an internal airline radio line.

"I was certain nobody could make a false call like that. But if somebody was blooming a weather balloon or something else over O'Hare, we were obliged to stop it since it was in very close propinquity to our flight working."

Eventually, most media outlets and the bigger one picked the story making people reappraise the idea of U.F.O.s. Despite the incident, investigations were turned down through the Federal Aviation Administration and the Airline. The F.A.A. ruled out the occurrence as a weather phenomenon and Dr. Mark Hammergren, who was an astronomer at Adler Planetarium, concurred, lamenting that the weather conditions at O'Hare that day were unfavorable.

"It's something that comes about when a propeller or jet airplane goes past when you have invariable cloud cover, and the temperature is right near the numbing point," Hammergren described. "They make liquid water droplets ice, and a foggy disc of ice crystals comes down from a hole. This makes it look like the cloud has been smashed, forming a perfect hole."

But for Director Mark Rodeghier, from the Center for U.F.O. Scientific Studies, the feasible alien spacecraft that materialized in the center's own conventional terrace remains a mystery worth investigating. "It is a mysterious object over O'Hare, and it's seen by official employees. Does United or the Federal Aviation Authority take it seriously? Of course, they don't. They have no interest because U.F.O.s do not exist. But how can you fail to be worried about something levitating over an airport after 9/11? This is not making any sense." Rodeghier remarked.

Journalist Leslie Kean talked about the O'Hare incident in her book U.F.O.s: Generals, Pilots, and Government Officials Go On The Record and even talked about it on "The Colbert Report" with Stephen Colbert in 2010, claiming that the government should probe the O'Hare case.

"This thing was levitating over Chicago O'Hare Airport at rush hour, a lot of people saw it ... the United States government did not say a word about it."

Accounts vary slightly, but everyone who witnessed this sighting seems to agree that the silent flying object sat in the sky for several minutes just before sunset. While several United employees swear they saw something anomalous, and in extension, one reported to have experienced some religious issues, the official Federal Aviation Authority maintains that the sightings were caused as a result of a weather phenomenon.

"That night was an exquisite climatic condition in terms of low cloud ceiling with a lot of airport lights, but when the lights glisten up into the clouds, you may see funny things sometimes." F.A.A. spokeswoman Elizabeth Isham was quoted saying.

O'Hare controller and union official Craig Burzych, who didn't spot the U.F.O., did spy a little bit of humor: "To fly 7 million light-years to O'Hare and then turning around because your gate was inhabited in utmost inadmissible."

U.S.S. Theodore Roosevelt U.F.O. Sighting

Following documents that were recently released by the Navy, fighter pilots reported close confrontations with unidentified aerial vehicles, as well as several critically close incidents, in eight occurrences between June 27, 2013, and Feb. 13, 2019. Two incidents took place in one day, according to one of eight unspecified Navy safety reports, which were disclosed in response to appeals filed under the Freedom of Information Act by media outlets.

In 2019, three videos of aerial encounters were authenticated by the Defense Department and were previously published by The Times, accompanying reports of Navy pilots who filed such close experiences. The encounters in the videos were probed by a little-known Pentagon project that for a number of years, investigated the reports on unidentified flying objects, and the Advanced Aerospace Threat Identification Program

While some of these happenings have been reported overtly before, the Freedom of Information Act revelations include the Navy's actual records that chronicled the incidents with the interpretation from the pilots on what they spotted. The Navy files, known as "hazard reports," recount both visual and radar sightings, together with the close calls they had with the aerial vehicles or "unmanned aircraft structures."

In late March of 2014, over the Atlantic Ocean off Virginia Beach, an incident occurred that entangled a silver thing "approximately the size of a suitcase" that was trailed on radar crossing within 1,000 feet of one of the fighter jets, according to a statement.

Some of the episodes involved fighter squadrons in the aircraft carrier, Theodore Roosevelt. One of the one-time F/A-18 Super Hornet pilots, Lt. Ryan Graves, reported a close encounter off Virginia Beach with what resembled a flying sphere caging a cube, as reported by a fellow pilot, and later narrated and filed with the squadron safety official.

The incident was cataloged in a report with few details on the 27th of June 2013, which declared that the Navy jet team saw something that passed about 200 feet away on the right side. With a clear smoke feather coming from the rear compartment, "the aircraft was white and was roughly the size and shape of a projectile or a drone," part of the report read.

No other organizations were supervising drone flights or launching missiles in the area at that time. This was according to the report. "Uncontrolled aerial vehicles exhibit a consequential mid-air clash commination," the commanding officer outlined.

The incidents included more than just that fleet, the VFA-11 "Red Rippers" out of Naval Air Station Oceana. The files show that the controllers took the incidents with seriousness, warning of the possibility of a mid-air crash. Defense Department officers do not represent the objects as extraterrestrial, and experts stress that earthly expositions can principally be found for such occurrences. Even lacking a credible terrestrial description does not make a space being one plausible, astrophysicists observe.

During the interviews, five of the pilots that were involved with the sightings avoided hypothesizing on the origin of the objects. The Navy, in its submissions, also avoided any such assumptions. Three episodes happened within exclusive U.S. airspace, meaning no other aircraft were directed to fly in that region. Another account of an incident that took place on 18th Nov. 2013 conveyed Panic. "With their small sizes, many U.A.S.'s are barely visually notable on radar manifest and therefore pose a serious risk for a mid-air crash," the report said, using an abbreviation for uncontrolled aircraft systems.

Experienced pilots in the U.S. Navy came across a number of horrifying encounters with U.F.O.s during training missions in the U.S between 2014 and 2015. While pilots were mid-air, their aircraft cameras and radar picked superficially impossible objects flying at supersonic speeds at altitudes beyond 30,000 feet. These strange U.F.O.s did so with no clear means of thrust. However, none of the pilots propose that these confounding U.F.O.s signify an extraterrestrial takeover.

In total, six pilots were responsible for the aircraft carrier U.S.S. Theodore Roosevelt in 2014 and 2015 told The Times about catching sight of U.F.O.s during flights through the southeastern coast of the United States, stretching from Florida to Virginia.

The two pilots who spoke with the publisher about the mystifying occurrences shared their experiences with the new History Channel documentary series "Unidentified: Inside America's U.F.O. Investigation." Videos of two aerial incidents emerge in the series, with clips of U.F.O.s: one-minute white speck and one large, dark bead. These U.F.O.s came to be known sequentially as "Gimbal" and "Go Fast."

The objects had "no clearwing, distinct tail, visible exhaust plume," Lieutenant. Danny Accion, a Navy pilot who publicized U.F.O. happenings since 2014, said in the film.

"It occurred as if they were conscious of our presence since they would resolutely move around us," Lt. Accoin explained.

According to the Lieutenant, when a mysterious reading comes up on the radar for the first time, it's practicable to interpret it as a false alarm, "but then when you start getting multiple sensors scanning the exact same thing, and then you get to see an array, that solidifies it for me."

Chapter 7: Recent UFO Investigations and Developments

In May 2020, the former President Barack Obama talked about the subject, saying there existed U.F.O.s whose nature is inexplicable. "What is true, and I'm actually being serious here, is that there's footage and records of objects in the skies, that we don't know exactly what they are, we can't explain how they moved, their trajectory. They did not have an easily explainable pattern. And so, you know I think that people still take seriously trying to investigate and figure out what that is"

To the Stars Academy

Tom DeLonge, who is a former Blink-182 guitarist, has implied that aliens may have been present at the birth of Jesus. The musician has for a while had interest in U.F.O.s, which made him co-found the company 'To the Stars Academy of Arts & Science.' The company works with U.S. government officials with the aim of "changing the world" through aerospace, entertainment, and other scientific pursuits.

During an interview with The Guardian, DeLonge claimed that there is evidence to back up his theory that extraterrestrials have been around since the beginning of humanity and suggested that the Bethlehem Star could have been an alien spacecraft. "These things were written in text thousands of years ago, things like hearing the voices in your head; like a burning bush that spoke to Moses," he described.

"The early texts may have called it God, but I'm saying it may not be that simple. The star of Bethlehem? Could it be a star or a spacecraft? You need to imagine how a star is that big. It definitely could not be hovering above a manger..." He added. In the same interview, DeLonge suggested that earlier, the U.S. government had considered sharing evidence of alien life but was afraid that people would not be able to take it lightly.

"I know there have been moments when certain presidents have come close to these aliens. The issue always is: how are people going to relate to these? How are they going to believe this? What will they think of the government that has hidden such information from them? The question will be, what more is there to know? It is scary at the Pentagon when they are trying to keep edification under the radar," explained DeLonge.

In early 2020, three videos that were thought to be U.F.O.s and that had already been publicized by DeLonge's own research organization gathered extra credibility after The Pentagon released them. The To The Stars organization published three clips in 2017 and 2018 that had been captured by Navy pilots, which showed strange objects that appeared to move rapidly within the United States airspace.

The New York Times reported the same footage; however, being the first release, the Pentagon was acknowledging its release, and their existence was reluctantly recognized by the Upper House. Pentagon spokesperson Sue Gough issued a statement saying that the videos had been officially released "to clear up any misconceptions by the public on whether or not the footage that has been circulating was real, or whether or not there is more to the videos." In September 2019, the

U.S. Navy also officially stated that the videos show footage of real "unknown" objects violating American airspace.

Luis Elizondo

According to Luis Elizondo, a former Pentagon official, he expresses his fears that findings on U.F.O. are being watered down by the powerful agencies within the government. In May of 2021, a month before the release of the U.F.O. classified information to the Congress, Mr. Elizondo told the New York Post that he would run for Congress to influence the Defense Department officials to make public all the findings on U.F.O.s, or U.A.P.s.

This followed a report that in The New York Times that alleged intelligence officials did not find any evidence of U.A.P.s, a month before the release of the report on 25 June, to the Congress. "If Public Affairs officers at the Pentagon's, and those who order them, keep obfuscating and misleading the American people about the truth on U.A.P.s, and what the government knows about them, I will have no option but to put myself out there and run for office, If that is what it will take to get the truth out, that's what I'll do," Elizondo said to The Post.

In his Twitter account, Elizondo said that should the reports issued by the Navy Pilots on U.A.P.s be true, then whatever objects the pilots could have witnessed around the world were more hi-tech than any technologies known to the intelligence service on earth. "It is time to release the full report, videos, and data that the Pentagon has."

Officials from the Pentagon told The Times that many documented reports from the last two decades did not involve the United States' technology and that other occurrences were incomprehensible. These occurrences include three contacts with U.S. Navy personnel, seen in videos which were disclosed in March 2020 by the Pentagon, after they were initially exposed first in 2007 and the other in 2017.

In the videos that were leaked, a pilot said he witnessed an object that spun through the air and went against the wind and was quoted saying: "Look at that thing, dude...it is rotating!" Mr. Elizondo was earlier on in charge of the Advanced Aerospace Threat Identification Program, which was conventionally shut down while Obama was the President.

In 2017 on another investigative unit by The Times, called the Unidentified Aerial Phenomena Task Force, the Pentagon admitted to continued investigations on U.A.P.s sightings years after the AATIP was closed down. Congress called on the defense department to disclose the full report on U.A.P.s in December, following the increasing reports of sightings and national security concerns.

Harry Reid

Harry Reid (D-Nev.), who is a former Senate Majority Leader, commented on the U.F.O. report, which is scheduled for release, and said he was glad he spoke out concerning it. The report, to be delivered by the Office of the Director of National Intelligence and the Secretary of Defense, was essential as part of a solution included in a coronavirus stimulus package signed into law by United States' former President Donald Trump in 2020.

Reid was among the first lawmakers to call for a probe into U.F.O.s. He was also responsible for one of the first works of scrutiny in 2007 on extraterrestrials. "I expect they are going to be fairly vague with what they come up with," the former senator was quoted by The Post relating to the much-awaited report. "Currently, we do not have enough information to draw presumptions."

Officials who had received briefings on the U.F.O.s and U.A.P.s report told The New York Times that no evidence indicated that U.F.O.s are alien space crafts. The report showed that the collected sightings that took over 20 years on U.F.O.s were not aircraft or technologies owned by the United States. The report also rules that the United States government has not conducted any secret operations on aliens or extraterrestrials.

Concerns have been raised that the seen objects believed to be alien space crafts could come from Russia and China, with the likelihood that the countries are conducting experiments with supersonic technologies.

Reid said the United States government should not stop investigations and rather challenged the government to establish whether these sightings could be a threat to national security. He also challenged the government to establish the nature of the technology used and whether it can be duplicated by the United States government.

The Pentagon's New U.F.O. Task Force

The former U.S. deputy defense secretary, David Norquist, established that the Pentagon has confirmed an Unidentified Aerial Phenomena (U.A.P.) Task Force (UAPTF) to revamp the understanding and gain apprehension into the nature and origins of Unidentified Aerial

Phenomena. According to a statement by the Pentagon, David Norquist approved the task force.

The Department of the Navy, under the apprehension of the Office Secretary of Defense for Intelligence and Security, led the UAPTF, whose objective was to detect, analyze and catalog U.A.P.s that could possibly become a matter of national security to the United States government.

"As D.O.D. has stated earlier, the safety of our people and the security of our undertakings are of paramount consideration. The Department of Defense and the military divisions take any attacks by uncertified aircraft into our training ranges or classified airspace very seriously and examine each information," the report stated.

"That incorporates the examinations of attacks that are initially reported as U.A.P. when the reporter cannot immediately recognize what he or she is observing," it added.

"The creation of a task force on Unidentified Flying Objects is another welcome move in the recent resumed interest and attention to these reports by government agencies and political players," said Mark Rodeghier, who is the president and scientific director of the J. Allen Hynek Center for U.F.O. Studies. "Without more details, it's unthinkable to judge how well-established the task force will be to seriously probe reports," Rodeghier added, "but I remain circumspect positive for the time being."

Rodeghier says he appreciates the need for secrecy. Nonetheless, "I would hope that as much data as possible is disclosed to the public so we can all be informed on this imaginably world-shattering matter," he is quoted from the interview he had with Inside Outer Scope.

Should D.O.D. Be Concerned?

"There are no doubts that military intelligence services around the world have always been fascinated with 'U.F.O. findings, there are real, unexplainable phenomenon behind them or not." These are views expressed by Jim Oberg, a space journalist, historian, and a controversial explorer of a slew of U.F.O. occurrences.

He's an acknowledged "lifelong space nut" and professional rocket scientist whose profession includes more than 20 years at NASA's Johnson Space Center. There are several reasons why the Defense Department is obsessed with U.F.O. findings:

• To recognize and improve instrumental irregularities in new technology, to make sure they don't, by mistake, misinterpret or overlook future findings.

• To establish how detection of irregularities might be intentionally induced by hackers and real enemies, and what can be done to block such attempts.

• To intentionally induce inconsistent targets into the range of new detection/tracking tech as a way of experimenting with them.

• To test enemy detection systems to identify exploitable weaknesses.

- To assess which findings from within or near opposing nations are indicators of their classified military testing and operations that they need intelligence of.

- To establish which detections, whether within or abroad, coincidentally reveal highly classified operations of our own that might be disclosed to enemy nations that are also looking for such things so as to improve our disguise, misdirection, and secrecy.

In so far as our domestic U.F.O. findings may be credible indicators of classified military technology, to purposefully create concealment and masking reports to deflect, confuse or quiet foreign observers and auditors.

"Insightful observers of the U.F.O. scene over the last two-thirds of a century have noted a significant feature of the evolution of reports. Their character has been changing, retaining eerie pace with the advancement in human observation and spy techs." Oberg says.

He adds that each year, the "tales on U.F.O.s" fade away just prior to the discovery of a new technology that would have explicitly documented them come online and be replaced by a new flavor of "peculiarities" that precisely match the restraints of new technologies.

Are Reports Open-Ended?

"I don't suppose this task force is as important as some people are purporting," says writer and U.F.O. pessimist Robert Sheaffer. "It's just a retaliation to all the publicity generated by 'To the Stars' exposing the three Navy infrared videos, which the Pentagon later published."

"At the military, a task force is something that is put together to handle a specific circumstance or problem," Sheaffer said. The military is then expected to submit a report and proposals about that issue and is shut down when such work is over. ""So this is not something open-ended and ongoing, like Project Blue Book. It does not suggest an ongoing government interest in unidentified objects," Sheaffer remarks. Carried out by the United States Air Force, Project Blue Book confirmed the existence of U.F.O. events starting in 1952. In 1970, the task force closed down.

Military Operation Areas (M.O.A.s) are clearly pointed out on aviation maps and non-combatant aircraft and are generally presumed to avoid them, Sheaffer explains. Most of the recent Pentagon sentiments about "unidentified objects" introduce "range invasions," the unidentified objects that seem to be entering one of these Military Operation Areas.

Hence, it seems that the military is concerned about the unknown objects that might be meddling within their sandbox. If the unknown objects turn up elsewhere, the military will not care. The 'Tic Tac' and 'Gimbal' videos seem to show far-off jets, which are probably well outside the Military Operations Area. The military is probing out of caution.

The creation of a task force to probe and evaluate Unidentified Ariel Phenomena is sensible. If it is done comprehensively, scientifically, and diligently, it can yield data that is useful in interpreting pilots' sightings.

Asteroid Oumuamua

In October 2017, a minute bright speck on the telescope at Haleakalā Observatory, Hawaii emerged from the celestial void toppling through space at 57,000mph. The object was thought to have come from the direction of Vega, an alien star, 147 trillion miles from earth. Shaped like a lengthened cigar, by the time it was identified it had already streaked by our own Sun, conducted a slick hairpin turn, and begun speeding off into a different direction.

This object was named 'Oumuamua', a Hawaiian term for "a messenger from far arriving first". This was not an ordinary comet or asteroid, it was an astral visitor from afar, unknown solar system, the first that was ever discovered.

There are many aspects of this object that make it strange and unique and two things specifically obsessed scientists. The first was its strange acceleration away from the Sun, which was difficult to relate with other ideas on what it might have been created from. The second was its unfamiliar shape which was 10 times as long as its width.

Before Oumuamua, the longest known space objects were three times longer than their width. Over the preceding years, scientific reports and global media headlines have been jammed with suppositions. Many people wondered whether it was a block of solid hydrogen, a cosmic dust bunny, or could it have been an artificial construction made by an intelligent extra-terrestrial civilization?

For now, we don't have any definitive answers. But speculation continues.

Chapter 8: The June 2021 Report

In June 2020, Senator Marco Rubio added some information into the 2021 Intelligence Authorization Act requesting that the Director of National Intelligence in-concert with the Secretary of Defense, submit "a comprehensive report on U.A.P. data and reconnaissance reporting."

This allowed the addressed a hundred and eighty days to submit the report. Mellon's proposals largely contributed to Senator Rubio's inquest, making it clear that this collaborative effort was more productive and more cost-effective.

In August of 2020, the Deputy Secretary of Defense, David Norquist, publicly announced the existence of the Unidentified Aerial Phenomena Task Force and were tasked to develop a report which they would present to the Congress on U.F.O. The Intelligence Authorization Act governing the Task Force was passed in December.

The June 2021 Report

Further scrutiny presented as evidence may not shed light on the causes of these sightings as they lack the level of strength required scientifically. However, new data that has been scientifically acquired may be useful and hence improve the understanding of the U.F.O. problem.

In June of 2021, a report is expected to be delivered to the United States Senate on the subject of Unidentified Aerial Phenomena, formerly known as Unidentified Flying objects or U.F.O.s. The motive of the report is to provide the lawmakers with data possessed by the Pentagon and the surveillance communities on incidents that seem to associate vehicles with startling flight traits that are way beyond the known advanced aircraft.

Earlier, New York Times published an excerpt on U.F.O. sightings hinting that details could still be scanty despite the reports being presented, and maybe this time around, again, there could be a cover-up in the report. The excerpt reads:

"American intelligence officials have found no evidence that aerial phenomena witnessed by Navy pilots in recent years are alien spacecraft, but they still cannot explain the unusual movements that have mystified scientists and the military, according to senior administration officials briefed on the findings of a highly anticipated government report."

However, CNN, in their report, in the follow-up of Times issue, published that:

"U.S. intelligence officials have found no evidence confirming that unidentified flying objects encountered by U.S. Navy pilots in recent years were alien spacecraft but also have not reached a definitive assessment as to what these mysterious objects might be, according to five sources familiar with the findings of an upcoming report on U.F.O.s that is expected to be delivered to Congress later this month."

Despite these reports, there's no indication that rules these as possible alien space crafts.

What Will The Report Find?

In theory, the report is expected to bring some light and resolve a thirty-year-old U.F.O. mystery of Nevada origins associated with Bob Lazar, on whether he worked with the Air Force, and specifically on whether he attended to an exotic craft that had been captured and secretly kept at a government facility in Lincoln County, near Area 51, called the S4.

In 1989, Lazar resurfaced with the pseudonym "Dennis" and was interviewed by George Knapp in Las Vegas. In the beginning, he spoke only in delineation; later, he came forward under his real name and with no impersonation. Lazar's claims were phenomenal. He stated that the U.S government possessed nine crashed spacecraft from outside planets. Amongst his claims, he described one spacecraft to have shaped like a disc.

According to Lazar, he had been part of a team that the government had hired to reverse-engineer the spacecraft, which would unlock the propulsion secrets that the American scientist needed to pave a path for the asteroids.

After bringing some friends into the desert to watch a saucer being tested near Area 51, Lazar was fired from his job at the clandestine military base. The group that Lazar had brought in was caught by a Lincoln County deputy who exposed him to the governing authorities.

Like other compelling elements of alien abduction, Lazar's story exposed the shadowed conspiracies that the government played under. His theory had an adverse influence on popular culture ranging from cartoons to movies.

The report that the Director of National Intelligence and the Secretary of Defense with other agencies are expected to present may include a detailed analysis of unidentified aerial phenomena data, which have been collected and are held by the office of Naval intelligence.

This data in possession of Unidentified Aerial Phenomena Task Force will contain data from sources like geospatial intelligence, human and signal intelligence, and measurements intelligence. A comprehensive report from the F.B.I., obtained from U.A.P. trespasses data over restricted U.S. airspace will be included.

The report is also expected to have identification of probable aerospace or other threats caused by the unidentified aerial phenomena to national security. In connection with this, it will contain an assessment of whether this U.A.P. activity may be credited to one or more foreign enemies.

Confirmation and approval of any incidents or motifs that indicate a potential rival may have achieved breakthrough aerospace aptness that could place United States strategic or conventional forces in danger.

Therefore, the Senate report has the capability of painting Lazar as a bigoted slanderer, an astral whistleblower with more influence than Edward Snowden, Karen Silkwood, and Daniel Ellsberg combined. The report may conclude that he is a liar if the reports submitted before the senate will overrule his story.

On the other hand, should Lazar be telling the truth, it is assumed the Pentagon will admit that there UFOs exist and will hence prompt the discussion of what is causing them. However, whether his arguments stand out as they are or get vilified, it is certain that we have alien hi-tech amongst us, which is capable of doing the unimaginable, as shown in the videos released and certified by the C.I.A. in 2019.

The Pentagon and U.F.O.s

Congress Confirms Videos and Photos of UFOs

In April 2021, the Pentagon confirmed that the leaked images and video of pyramid-shaped unidentified flying objects were real and were captured in 2019 by a U.S. Navy Pilot. According to the then Pentagon's spokeswoman Sue Gough, the grainy video captured in night vision, together with three other photos of unidentified aerial phenomena, were taken by the Navy's Snoopy special team flying over the USS Russell.

"Like we have said before, to maintain security operations and to avoid releasing information that may be helpful to potential opponents, DOD does not deliberate publicly the details of either the findings of the examinations of reported attacks into our training ranges or classified airspace, including those attacks initially classified as Unidentified Aerial Phenomena," Gough said while issuing a statement on UFOs.

The following month, May 2021, Gough confirmed another footage that was taken aboard the USS Omaha in 2019 off the coast of San Diego in California. Gough said, "I can confirm to you that the video was taken by the Navy crew, and it has been included in the ongoing investigation."

The video, which was released by Filmmaker Jeremy Corbell, showed the unidentified craft entering the ocean without crashing. The Pentagon referred to these crafts as "trans medium vehicles," to mean that they moved through different mediums like air, vacuum space, and water.

Growing Public Belief

In "U.F.O.s: Generals, Pilots, and Government Officials Go on the Record," by Leslie Kean, and which was published in 2010, she wrote that "the U.S. government routinely ignores U.F.O.s and, when pressed, the government issues false statements. Its indifference and dismissals are incautious, disrespectful to dependable, often expert witnesses, and potentially treacherous."

However, since the Second World War, almost half of Americans have welcomed the idea of U.F.O.s and have been increasingly interested in the topic. In her book, Kean considers herself as a watchdog of this lost history.

The issue of how vital it was to treat and consider what they renamed UFOs evoked a deep conflict within the government. By September of 1947, filed reports of sightings had become too profuse for the Air Force to disregard. That month, through classified communications, General Nathan F. Twining counseled the commanding general, saying that "the phenomenon reported was something real and not intuitive or apocryphal."

The "Twining memo," which gained a canonical reputation among ufologists, highlighted concerns that some foreign rivals, at this point the Soviet Union, had made an inconceivable technological breakthrough.

Officials were equally split between those who believed that the "flying discs" were without a doubt, "interplanetary" in origin and those who pinpointed the disclosure to rampant astigmatism. According to a memo that was given, on the one hand, a whole twenty percent of U.F.O. sightings lacked regular explanations while on the other hand, there was no conclusive evidence like the debris or fragments of a crashed saucer, as a scientist at the RAND Corporation thought, astral travel was simply absurd.

The habit of not considering, disapproving, or overlooking unseasonable facts is a thing the people who are against and those who believe in something have in common. One British explorer has compellingly shown that the Rendlesham case, commonly referred to as Britain's Roswell, may have been merely a meteor, a lighthouse that could be seen through the woods and fog, and the preternatural noises made by a barking deer.

Reports filed by eyewitnesses are apt for considerable embellishment over time, and strings of coincidences can easily be proffered into a meaningful pattern by a human mind prone to misinterpretation and anxious for confirmation.

The explorer had comprehensively debunked the claim, but Kean seemed unperturbed by his verdict. Kean's idea suggests that such accounts, as Rendlesham described, violated Occam's razor. Even if Rendlesham was "complicated," she wrote, it was still "one of the top ten U.F.O. sightings ever recorded."

There were always other occurrences. "The U.F.O. Experience," as written by Hynek, had asserted that U.F.O. sightings represented a phenomenon that needed to be taken into consideration. Following more reports, it was thought, hundreds upon hundreds of outstanding stories told by trusted people were to be put together.

Practically, all astrobiologists are skeptical that we are alone. According to the senior astronomer at the SETI Institute, Seth Shostak, he has pledged that, by the year 2036, we will find irrefutable proof of astute life. Astronomers have deliberated that there may be hundreds of millions of possibly inhabitable exoplanets in our cluster.

Astral travel by living beings still seems like a strange and insignificant possibility. However, since the early nineteen-nineties, physicists have known that faster-than-light travel is plausible in theory, and new research has brought this narrowly closer to being attainable in practice.

These approaches, as well as the inference that our civilization may be one that could be millions and/or billions of years behind our galactic neighbors leave open the very real possibility that extraterrestrials could be around us. They have certainly made their presence visible, hence proving the idea that U.F.O.s have extraterrestrial origins.

The government may or may not cooperate in the unveiling of the U.F.O. mystery. But, in throwing up its hands and acknowledging that there are things it cannot understand, it has loosened its grip on the narrative.

For the majority, this has been a consolation. For all the usual reasons, governments have never officially reported the sightings nor informed the public that these extraterrestrials exist, but the report that is due for release is making people excited for the truth after all these years.

Conclusion

Over the years, researchers and scientists have combed government documents at the National Archives in search of evidence of life's existence beyond earth. In an independent analysis of UFO occurrences since 1970, scientists have determined that some sightings go along with physical evidence that warrants a scientific review. However, there is no proof that points to a breach of known natural laws or the involvement of extraterrestrial beings.

The National Archives and Records Administration stores a cluster of documents that have relevance to unidentified flying objects (UFOs) or "flying disks." Over the decades, those assets have been thoroughly investigated and scrutinized for hints of more information and evidence that aliens exist.

Project Blue Book, which includes retired, known records from the United States Air Force (USAF), is in the protection of the National Archives. It is concerned with the 1947 to 1969 USAF investigations on UFOs.

A US Air Force Fact Sheet reports a total of 12,618 sightings which were reported to Project Blue Book in the course of 1947 to 1969. Seven hundred and one were unknown. Richard Peuser, chief of literal credential operations at the National Archives in College Park, Maryland, said the organization saw a stable amount of interest in files that handled UFOs, responding to "a few hundred requests" over the time.

"Sometimes the same people would send us several emails expecting that they would receive different feedback," Peuser continued. "Many felt that the reports were too congenial and that the Government was concealing the truth from them. There were often assertions of coverup, of deliberately keeping or sabotaging the reports."

According to Peuser, they still get a fair number of inquests that are UFO-related, and some people have even come in searching for other files in elevation, specifically in the US Air Force reports. Therefore, Roswell, Area 51, Majestic-12, Projects Mogul, Sign, Grudge, and Twinkle keep fascinating and drawing researchers to examine resources on extraterrestrials.

The National Archives inventory yielded 37 catalog explanations, assembled under flying saucers, flying UFO phenomena, UFOlogy, or UFOs. As records have been processed and systemized, other documents have resurfaced.

A few years ago, technician Michael Rhodes was processing boxes of Air Force reports when he stumbled upon a drawing in the corner of a test flight document that caught his attention. The drawing had conspicuous correspondence in detail to the flying saucers in popular science fiction films made during that time.

The Archives online inventory includes a series of reports and documents from the Federal Aviation Administration that report the sighting of a UFO by the team of Japan Airlines Flight 1628 while in Alaskan air. The reports include simulated radar representation and an article that featured in The Philadelphia Inquirer Magazine on May 24, 1987, about the occurrence.

Records in this assemblage also contain notes from interviews with the three crew members who saw the UFO and have been made available in the online inventory. It was discovered that the records were part of the Alaskan digitalizing project, as the thought of by a metadata technician at the National Archives in Seattle, Washington, known as Marie Brindo-Vas.

The other fascinating record from the National Aeronautics and Space Administration files is the Air-to-Ground Gemini VII documentation. This was found in the online catalog, and the records involve the transcript of communication between the Houston control center and astronauts who "at 10 o'clock high have a bogey."

Bogey was the code used to mention the Unidentified Flying Objects. The communication goes on to elucidate that the astronauts are spotting in a polar orbit "hundreds of little objects going by from the left at about 3 to 4 miles out."

The Archives also has audio records relating to UFOs, such as the video of Maj. Gen. John A. Samford's pronouncement on "Flying Saucers" from the Pentagon, Washington, DC, in July 1952, which the military leader talks about the Army's probe of flying saucers. Another clip issued by the Department of Defense shows the United States Air Force Lieutenant colonel Lawrence J. Tacker and Maj. Hector Quintanilla, Junior., discussing Project Blue Book and the discerning of Unidentified Flying Objects.

The Gerald Ford Presidential Library and Museum have a report that relates to UFOs, and which was created by Ford when he was the House Minority Leader and Congressman. The original report is located in Box

D9, folder "Ford Press Releases – Unidentified Flying Objects, 1966" of the Ford Congressional documents: Press Secretary and File.

In this epistle, then-Congressman Ford suggested that "Congress probe the rash of reported occurrences of UFOs in the United States." An accompanying news release to that memo goes on to say, "Ford is not compelled with the Air Force description of the recent sightings in Michigan and cites the "swamp gas" version given by astrophysicist J. Allen Hynek, as frivolous."

In October 1969, Jimmy Carter, the then-Governor of Georgia, spotted a UFO over the skies of Leary in Georgia. The Jimmy Carter Presidential Museum has the full account that he handed over to the International UFO Secretaire.

As more documents are searched, processed, and revealed, what evidence might be found of alien and UFO existence throughout the libraries in the United States? That remains to be unknown. However, based on past history, it's believable that researchers and UFO fanatics will keep digging for more data and reports.

The extensive fascination with the likelihood of the existence of extraterrestrial beings, forms, and UFOs continues to create great passion and controversy all over the world.

Despite reports on UFOs dating back to 50 years ago, the information collected does not confirm that either unknown physical processes or extraterrestrial technologies are compromised. It, however, concludes that a deep investigation may be important to keenly assess UFO records in order to extract information about atypical phenomena that only science can define. In order to be feasible to the scientific community, such assessments must take place with a spirit of impartiality and a willingness to appraise rival suppositions that have so far been missing.

Further analysis of the proof submitted by the scientists and researchers is unlikely to shed added light on the reports. Most current Unidentified Flying Objects investigations do not have the level of strength required by the scientific community, although it involves the initiative and commitment of the investigators. However, new data that has scientifically been acquired and reviewed may produce helpful information and advance our perception of the UFO phenomena.

References

1. Fontaine, Frances. *Reader's Digest Mysteries of the Unexplained.* Pleasantville, NY: Reader's Digest Association, 1985. *UFO Sightings in Ancient Egypt, Rome, And the Middle Ages.* Rense.
2. Maloney, Mack. *UFOs in Wartime: What They Didn't Want You to Know.* New York: Berkley, 2011. Print.
3. Monfort, Sam. "I Want to Believe: UFO Sightings Around the World." *Visualize This.* Visualize This, 22 Feb. 2017. Web. https://vizthis.wordpress.com/2017/02/21/i-want-to-believe-ufo-sightings-around-the-world/
4. Stothers, Richard. "Unidentified Flying Objects In Classical Antiquity." *The Classical Journal* 103.1 (2007): 79-92. *NASA Goddard Institute for Space Studies.* NASA.
5. Records on flying saucers from the Special Subcommittee on Government Information, Committee on Government Operations, 86th Congress; Records of the U.S. House of Representatives, Record Group 233; National Archives, Washington, D.C.
6. Web. http://www.rense.com/general7/ages.htm
7. Web. https://pubs.giss.nasa.gov/docs/2007/2007_Stothers_st02710y.pdf
8. Fontaine, Frances. *Reader's Digest Mysteries of the Unexplained.* Pleasantville, NY: Reader's Digest
9. Association, 1985. *UFO Sightings in Ancient Egypt, Rome, And the Middle Ages.* Rense.
10. Anatomy of a phenomena, J. Vallee, Ace Books, Inc H-17
11. The Report of Unidentified Flying Objects, E.J. Ruppelt, Ace. Books, Inc. G-537
12. Report on UFO Wave of 1947, Ted Bloecher; 1967

UFO Report UPDATE – July 2021

Background

On June 25th, 2021, a new report titled 'Preliminary Assessment: Unidentified Aerial Phenomena' was published by the 'Office of the Director of National Intelligence' of the United States. This report was dubbed by both the mass media and the public at large as the "UFO Pentagon Report," although the department itself refers to the report as the UAP Report. It is unusual that the United States government has taken to referring to such phenomena as "unidentified aerial phenomena" or "UAP," as opposed to the far more conventional "unidentified flying objects" or "UFOs." This may be to distract the general public, or to dissuade them from associating these unidentified objects in the sky with extra-terrestrial or even extradimensional life. Or it could simply be that they find the term "unidentified aerial phenomena" to be more accurate and specific. Maybe "flying" is too specific itself and implies too much in the name, so they went with something broader. At this point, it still remains unclear why they insisted on this alternative title.

Regardless of the reason for the adoption of this unusual nomenclature, this assessment was federally mandated and has stoked a lot of popular interest with the general public. In fact, 43% of the population of the United States are reportedly interested in the phenomena of unidentified flying objects (or unidentified aerial phenomena), and many United States officials and politicians have had their say on the issue, including Barack Obama, who is reported as saying, "there's footage and records of objects in the skies that we don't know exactly what they are," and Marco Rubio, who has said that "there's stuff flying in our airspace and we don't know who it is and it's not ours."

The report was funded as part of the 'Consolidated Appropriations Act 2021', a $2.3 trillion dollar bill that was ostensibly passed for Covid 19 relief. It was intended to collate information from a variety of governmental departments, including the FBI, the Unidentified Aerial Phenomena Task Force, and the Office of Naval Intelligence. The Unidentified Aerial Phenomena Task Force itself was set up for the very purpose of monitoring what are conventionally better known as UFOs, beginning in 2020 as a successor to the Advanced Aerospace Threat Identification Program, which began in 2007 and was closed in 2012. It was the Department of Defense's release of the three videos now known as the "Pentagon UFO videos" that stoked public interest and led to the initiation of this particular report.

Since its inception, the Advanced Aerospace Threat Identification Program has received over twenty-two million dollars in funding, further stoking public interest. After all, if there were little or no expectations of extra-terrestrial life to be discovered, why would the United States government fund the project with so much money? After the establishment of the Advanced Aerospace Threat Identification Program, one of the main program directors, Luis Elizondo, founded the 'To the Stars Academy of Arts and Science.' This academy is a Las Vegas-based public-benefit corporation that was co-founded by the famous parapsychologist Harold E. Puthoff (known for his involvement with the Church of Scientology and for practicing and promoting such psychic phenomena as remote viewing and extrasensory perception), and, bizarrely, Tom DeLonge, a famous musician known for founding and performing in Blink182 and Angels & Airwaves. Curiously, the corporation began as a music label and publisher, and their works include the fifth Angels & Airwaves album, 'The Dream Walker', and Tom DeLonge's aptly-titled solo album, 'To the Stars'. They later developed into a research corporation investigating remote viewing, ESP, and, indeed, UFOlogy. It is curious that what might well be considered a fringe parapsychological organization is so heavily involved and overlapped with the US military and the Advanced Aerospace Threat Identification Program.

What the Report Says

The report opens with a summary of the scope and features of the project. The basic background is offered first (as summarized in the above section) and goes on to explain that the report is intended as an overview of the challenges associated with the overwhelming evidence and material on unidentified aerial phenomena. It goes on to say that the purpose of the report is to explain the difficulty in establishing appropriate methods, policies, and training to identify and deal with these still unidentified phenomena, to enhance the intelligence community's understanding of this threat, and to establish whether indeed it is a threat at all (although, considering the fact that by this time these unidentified aerial phenomena have caused numerous military training exercises to be aborted prematurely, it is safe to say that they are *something* of a threat, at least so long as they remain unidentified). It then explains that the report refers specifically to phenomena observed by the United States military between 2004 and 2021.

In the 'assumptions' section, the authors of the report address the potential of sensors performing poorly. They argue that generally speaking, military-grade sensors designed to register unidentified aerial phenomena will typically perform and record accurately, although there is some chance of error in the form of sensor anomalies. The report then moves on to the executive summary section.

The executive summary section opens with the statement that "the limited amount of high-quality reporting on unidentified aerial phenomena (UAP) hampers [the United States government's] ability to draw firm conclusions about the nature or intent of UAP." This is because the data currently being analyzed comes from a variety of sources and is designed to record quite a broad range of phenomena, from saucer-like objects to aerial ball-shaped objects to other strange and inconsistent phenomena. Therefore, it is hard to reach general conclusions about any of the occurrences that were recorded and observed. Generalities must be reached (otherwise, the report would be pointless), but it is difficult to generalize when the source material is so varied. The reporting, with evidence taken from the United States military and the Intelligence Community, or IC, "lacked sufficient specificity," and therefore those producing the report found it very difficult to generalize about what was observed.

The report then goes on to explain that in some of the footage of unidentified aerial phenomena that was observed, some of the objects appear to make movements unknown to contemporary (Western or United States) science. Some of the objects move rapidly and change direction at too fast a frequency to understand, some of the objects appear to defy gravity by the sorts of movements they make, and others exhibit other forms of "unusual flight characteristics."

However, the report does note that these unusual flying characteristics may simply be the result of "sensor errors, spoofing, or observer misperception" and that they likely will require "additional rigorous analysis." Sometimes, what can appear as anti-gravity activity or other kinds of unorthodox movement have to do with sensor equipment or even the person behind the camera. For example, an unidentified object in a perfectly "blank" sky (just black or blue with no clouds or other points of references) can appear to move in strange ways if the camera is moved slightly in one direction or the other. The slightest nudge of the recording equipment can therefore make it appear as though the object itself has moved or changed course when in fact, if the camera had remained steady, it would have either remained static or stuck to its initial course. For this reason, the unusual kinds of movement exhibited by some of these unidentified objects may not necessarily be the result of movement from the object itself and therefore, the examination of this movement is currently inconclusive.

The report then goes on to explain that, given the different shapes as well as the different kinds of movement exhibited by the aerial phenomena being observed, that there may well actually be different kinds of objects from different sources being observed, rather than all being a part of the same "phenomena." In other words, one object could be a Russian satellite, another could be a deflating balloon, while another could be evidence of extraterrestrial life. It is not clear that all these phenomena belong within the same category or that they were all produced by the same source.

Then the report highlights the fact that the unidentified aerial phenomena do pose a very real threat to United States national security as well as the safety of air flight. An "increasingly cluttered air domain" is a threat to safety and a complication for various forms of flight for obvious reasons, especially as long as these objects and phenomena remain unidentified. It is therefore crucial that these phenomena are investigated fully and identified, whether they are a deliberate threat caused by an adverse power (terrestrial or otherwise) or are simply something causing a kind of accidental threat simply by their unannounced presence in United States air space. The executive summary then concludes by explaining that the consolidation and streamlining of all of these various reports of unidentified aerial phenomena is crucial for a deeper understanding of such phenomena.

The following section is titled 'Available Reporting Largely Inconclusive,' and explains that although we have a lot of great observatory material to work with, the findings and conclusions are still inconclusive with regard to what these phenomena actually are or, in other words, turning the unidentified aerial phenomena into identified aerial phenomena. It is explained that there was no standard reporting mechanism until the Navy developed one early on in 2019, and that this was only adopted by the Air Force in late 2020. 144 different objects or forms of aerial phenomena were established, with only one being identified with confidence. This turned out to be a large balloon that was deflating, which explains its unusual and unexpected movements. However, of the other 143 sightings and observations, they have not been reasonably explained or identified with any known United States technology. Further, 80 of the 144 were recorded by multiple forms of sensors. Therefore, for at least 80 of the observations, we can rule out the possibility of an anomalous sensor error. The limits of the data are the main reason that these phenomena remain unidentified and unexplained.

This section also explains that there is a stigma attached to reporting such phenomena within the military. Therefore, while eyewitness accounts of military personnel are important, there is the fear that some phenomena, especially that which resembles some kind of extraterrestrial life, are going unreported because the personnel in question do not wish to appear as "quacks" or be suspect in some other way. It is said that this stigma has decreased since the scientific and intelligence communities have been involved in these observations. Nevertheless, there is still a certain level of reputational risk that must be taken into consideration when we account for the limited data that we must draw conclusions from.

"Sensor vantage points," "optical sensors," and "radiofrequency sensors" also play a key role in these investigations as, unlike standard video recording, this kind of sensor equipment can actually help to determine the relative size, shape, and structure of the phenomena being observed.

Despite the limited data received, inconsistency of records, and potential for withheld information (due to stigma and reputational risk), some overall patterns in the data have emerged that make investigation a lot easier. The report goes on to explain: "*Although there was wide variability in the reports and the dataset is currently too limited to allow for detailed trend or pattern analysis, there was some clustering of UAP observations regarding shape, size, and, particularly, propulsion.*" It is noted that these phenomena tend to cluster around the training grounds of the United States military but that this is likely due to some kind of collection bias (similar phenomena may be occurring elsewhere, but not actually being recorded by military-grade sensor equipment).

This section of the report then moves on to explaining that much of the phenomena observed appeared to demonstrate "advanced technology." In 18 of the observations (taken from a total of 21 different reports), "observers reported unusual UAP movement patterns or flight characteristics," including the aforementioned shifts in direction or movements suggestive of anti-gravity technology. The report explains in more detail:

"*Some UAP appeared to remain stationary in winds aloft, move against the wind, maneuver abruptly, or move at considerable speed, without discernable means of propulsion. In a small number of cases, military aircraft systems processed radio frequency (RF) energy associated with UAP sightings.*"
This means that not only did these aerial phenomena exhibit signs of anti-gravity technology, but also anti-wind measures, whilst not showing any obvious signs of propulsion, at least not by means of any technology known to the United States military. The report then explains that further investigation is needed to determine whether or not "breakthrough technologies" have been discovered.

The next section of the report is titled, 'UAP Probably Lack a Single Explanation,' and, as you might expect, it goes into detail about the fact that the variety of aerial phenomena observed seem to imply different modes of transport and sources of technology, so there could be a variety of unrelated phenomena being observed.

It opens with a brief introduction to the subject, explaining that *"the (unidentified aerial phenomena) documented in this limited data set demonstrate an array of aerial behaviors, reinforcing the possibility that there are multiple types of UAP requiring different explanations."*

It then goes on to describe the "deflating balloon" conclusion reached for one of the objects (it is not the first time that a balloon deflating in the air has been mistaken for an extraterrestrial; the crash landing in Roswell is believed to have been a broken weather balloon). It then goes on to list various categories that the unidentified aerial phenomena might fall under. Airborne clutter includes everything from birds to balloons to plastic bags to "recreational unmanned aerial vehicles" (this would include personal drones, remote control plane toys, etc.), and so on. Natural atmospheric phenomena include moisture, ice crystals, "thermal fluctuations," and various other phenomena related to the weather and atmosphere. USG, or "Industry Developmental Programs," refers to classified projects and programs by the United States that the report was unable to confirm or deny, while foreign adversary systems refer to programs and projects by ambivalent or hostile world powers such as Russia, China, North Korea, and so on.

The final section, 'other,' is a kind of miscellaneous category to collect all of the phenomena currently unidentified. The report states that

"...most of the (unidentified aerial phenomena) described in our data set probably remain unidentified due to limited data or challenges to collection processing or analysis...we may require additional scientific knowledge to successfully collect on, analyze, and characterize some of them."

The next section goes into detail about the various ways in which unidentified aerial phenomena might threaten or compromise United States security and military activity, stating that *"(unidentified aerial phenomena) pose a hazard to safety of flight and could pose broader danger if some instances represent sophisticated collection against U.S. military activities by a foreign government or demonstrate a breakthrough aerospace technology by a potential adversary."*

These potential threats are then broken up into two further categories. The first is that of "ongoing airspace concerns," which covers all of the problems that might be caused for United States aircraft that have to change their course or compromise their behavior to avoid collision and other unwanted interactions, and the second is "potential national security challenges," which could be a problem if these phenomena really do represent spying or sabotage potentiality by hostile entities such as Russia, China, North Korea, and so on. To this should be added the possibility of hostile extraterrestrial powers, which are not referred to directly but could well be a very real threat.

The report then goes on to reiterate that explaining unidentified aerial phenomena will require broader analysis and a greater collection of data, as well as greater investment in resources designed specifically to monitor such phenomena. Specifically, the United States military seeks to "standardize the reporting, consolidate the data, and deepen the analysis" of unidentified aerial phenomena. Part of this expansion of reporting is to consolidate certain patterns. If, for example, there are numerous entries of a balloon deflating or a type of bird in flight, a consolidated pattern recognition process might be able to detect the familiar pattern associated with these phenomena and more accurately pinpoint their source before registering them as unidentified aerial phenomena.

In this section, it is also explained that most of the data collected has come from the United States navy, while future reports should collate data and observations from a broader range of United States military faculties, along with other governmental agencies. They also wish to expand the collection of data beyond military bases and thereby eliminate the collection bias in the dataset associated with taking so much data from one type of source (military training spaces). Finally, the report highlights the necessity of increasing investment in research and development as guided by the UAP Program Plan, UAP Collection Strategy, and UAP R&D Technical Roadmap.

The report then ends with Appendix A, being a glossary of key terms used throughout the report, and Appendix B, being a summary of the senate report accompanying the Intelligence Authorization Act for the fiscal year of 2021.

Conclusions and Questions

There are a number of initial takeaways that can be drawn from this report.

- **No explicit mention of extraterrestrial life:** First of all, it is intriguing to note that despite the popular interest in this investigation and report revolving around the possibility of extraterrestrial life making its presence felt on Earth, the report makes no mention at all of aliens, extraterrestrials, extra-dimensional beings, or anything involving the paranormal, parapsychological, or cryptozoological. In fact, the report doesn't even use the terms "UFO" or "unidentified flying objects."
- **Numerous categories of entities have been established:** The categories of possibility listed by the report include weather and atmosphere, recognizable flying objects such as plastic bags and wildlife, classified technology from the United States, and technology from hostile states such as Russia and China.
- **Unidentified aerial phenomena are considered to be a threat:** Unidentified aerial phenomena could be a threat to the United States for two reasons. First, they compromise aerial flight as several military exercises had to be aborted for reasons relating to these phenomena – in the future, this may well affect commercial flight alongside military flight. Second, if these strange aircraft are the product of hostile powers (terrestrial or extraterrestrial, although only terrestrial powers are referred to in the report) then this poses a threat to the United States for obvious reasons.
- **The unidentified aerial phenomena observed exhibit unusual and unexplained movement:** This unusual movement includes a rapid change of direction, propulsions with lack of a clear means, anti-gravity capabilities, anti-wind

capabilities, and the ability to remain in the air motionless. Although they are not listed in the report, footage of the unidentified aerial phenomena shows some of them dipping effortlessly in and out of the water (moving under the water with just as much ease as through the air) and one object apparently *splitting in half* and traveling off into opposite directions.

- **Most of the unidentified aerial phenomena were observed to appear around military bases:** The key here, however, is the term 'observed'. A potential and probable collection bias has been noted by the report, as all of the most sophisticated sensor equipment designed for this purpose is located at military bases. The report goes on to explain that they plan to remedy this by hosting sensor equipment away from military bases. This will help to establish whether they are indeed the result of collection and observational bias, or whether these strange aircraft are actually attracted to military bases for one reason or another.

- **Only one of the 144 objects observed has been confidently identified:** The only example of unidentified aerial phenomena being identified was that of a large, deflating balloon. While this is now considered "identified," there are still another 143 observations yet to be identified.

- **The unidentified aerial phenomena likely lack a single cause:** In other words, it is expected that there are probably multiple unrelated phenomena being observed. One sighting might simply be a bird or some other wildlife, another might be a piece of paper floating around, another might be some kind of satellite or spying equipment from China or Russia, while another might be an extraterrestrial aircraft (although the latter is, of course, downplayed in this report).

- **There is a stigma associated with the reporting of unidentified aerial phenomena:** While this has decreased somewhat with both the scientific community and popular opinion being interested in unidentified aerial phenomena, there is still a stigma and some degree of reputational risk associated with the reporting of unidentified aerial phenomena which may be a hindrance to future reporting.
- **Investigation of these unidentified aerial phenomena is ongoing:** So far, data has mostly been collected from the navy. The United States government plans to expand this investigation into other departments of the military and government and to streamline the process of pattern recognition in order to reach more conclusive judgments.

The United States government still maintains that the unidentified aerial phenomena recorded by the military are most likely to be a "foreign threat." The chair of the Senate Foreign Relations Committee was quoted as saying, "if something's out there, let's seek it out, and it's probably a foreign power." However, the possibility that this foreign threat is an *extraterrestrial* threat has not been dismissed by experts and observers. Boston University astronomer Thomas Bania has spoken at length of the possibility. In an interview with the Boston University website, he says, "do any of us believe these are actual flying saucers from outer space? No. But there really isn't enough information to evaluate what's going on." He talks about how the speed and sudden change in direction is unusual, but it is unclear just how fast these objects were moving without some idea of their approximate distance. He explains that, if we look at a regular commercial plane flying in the sky, given the distance that we are looking at it, it might appear to be moving very slowly. However, if we zoom in very close to the plane, it will appear to move very quickly (as it actually is). Similarly, it is not easy to gauge the speed of these strange flying objects without having any idea of our level of magnification. If we are zoomed in very closely, but believe that we are looking from a distance, the objects might give the illusion that they are moving a lot faster than they really are. *"The point is, how do you interpret what these experts are responding to—the enormous speeds and maneuvers,"* he explains, *"I am questioning that interpretation of the data; without knowing the distance, there would be no evidence of enormous accelerations and avionic capabilities in terms of maneuverability."*

Banla's colleague, Jack Weinstein (security expert at Boston University), is somewhat more open to the idea of the subject of this report actually being real UFOs. He explains:

"It could truly be a UFO. I never want to say something could never be...[or] it could just be a phenomenon caused by nature—just a weather phenomenon. We learn something new every day, we're flying at higher altitude than we've ever flown before, so maybe we're seeing things that would look differently at a lower altitude." He then expands on the interest and speculation of UFOs by the broader public, including conspiracy theorists: *"I think everyone who believes in UFOs will see a conspiracy if the government doesn't say they are UFOs. I can't believe we're the most intelligent life-form in the entire universe, 'cause that*

means the universe is pretty dumb. If I'm an extraterrestrial and I can travel from other planets, I would think they would be smart enough to evade radar."

Classified version

While the initial report is only nine pages long, one of the most interesting things about this investigation – especially for those interested in the prospect of contact with extraterrestrial life – is the fact that there is a longer, classified version of the report that has not been released to the public. Some sources claim that this report is nearly double the length of the version released to the public, at around 17 pages in length, while other sources have said that the total length of the uncensored report is closer to 70 pages in total. There has been much speculation as to the contents of this report and why at least half of the report might have been kept from the general public.

Advocates of UFO and extraterrestrial theories of unidentified aerial phenomena have pushed the idea that the content that remains classified contains evidence of extraterrestrial life. Perhaps the report was censored because they did not wish to alarm the public, or because further investigation is required. It has also been said that Area 51 was likely involved, and due to the fact that the activity at Area 51 is highly classified, that this is the reason that these details were omitted from the final report, or at least the redacted version that was released to the public.

On the other hand, skeptics of extra-terrestrial activity have pushed for an alternative theory, namely that the report refers to classified United

States technology, such as sophisticated spying equipment, and that the reason it was left out of the public version of the report is that it would otherwise leak into the hands of Russia, China, North Korea, and other hostile powers, thwarting the United States spying and security efforts and even handing over their technological secrets to these foreign powers.

An American historian named Richard Dolan has recently made the news by claiming that the classified version of the report – which he purports is 70 pages long in its total length – actually contains reports of the events and experiments taking place at Area 51, and that the report could, in all likelihood, contain content about experimental technology of adverse powers – perhaps hostile foreign powers, or perhaps even extra-terrestrial powers. Dolan claimed specifically that the Pentagon has been investigating *"ET-related items...energy pulse propulsion, ion propulsion, anti-gravity propulsion...[and]...anti-matter propulsion."*

Even previous presidents such as Bill Clinton and Barack Obama have made comments pertaining to the fact that something mysterious, something yet not understood, is taking place in the skies above us.

During the presidency of Donald Trump, the president reorganized the 'Air Force Space Command' into the 'United States Space Force' and then went on to establish this force into a new branch of the United States military, the most recent in fact since the Army Air Forces were reconfigured into the United States Air Force in the 1940s. While this reestablishment of the United States Air Force was ostensibly performed for 'security' reasons with hostile foreign powers in mind, it is unquestionable that such a force will be used to observe and investigate other space-based aerial phenomena, including UFOs. In response to the trumped-up space force, Putin has also made statements concerning this 21st-century space race, promising to ramp up funding for Russia's own space force as a counter to work already established by Donald Trump and the United States.

Another UFO expert – professor and researcher Bob McGwier – has claimed that the video footage shown during the briefing for the classified version of the report was like something from "a science fiction movie," and went on to claim that most of the important and interesting aspects of the report and investigation were removed from the version of the report that is available to the general public and that the classified version of the report contains far more intriguing content. Meanwhile, a United States congressman from the Democrat Party, Andre Carson, is on record as saying, "there have been nearly 150 sightings…80 of those sightings have been detected with some of the best technology the world has ever seen…we can't rule out something that's otherworldly…"

Summary

In sum, there is a good reason that the recent videos and pictures released by the US military, and the subsequent investigation and report into this footage, have stoked a lot of interest among enthusiasts of unidentified flying objects in the United States and around the world.

First of all, the footage itself is shocking even before the investigation began and the report was released. Some of the footage shows orb-like objects seemingly defying gravity, others show flying saucers that look exactly like the extra-terrestrial sightings of previous decades (to say nothing of popular fiction and sci-fi movies), there have been strange shapes moving quickly and changing direction that show no obvious signs of any means of propulsion (at least no means known to contemporary science), and we've even seen what appears to be brand new technology, such as one flying object which appeared to dip in and out of the ocean, moving through both atmospheres with ease, before splitting in half and taking off in two different directions.

Despite the sophistication of the United States military in the 21st century, these sightings still baffle investigators and researchers, and no firm conclusions have been reached regarding what these things are. There are various possible categories, such as wildlife, weather, and hostile governments, but none of these have been confirmed. Regardless of the source of this unidentified aerial phenomenon, it is clear that we are dealing with a form of technology currently unknown to the United States.

The United States government are intent on continuing investigation into these matters and even expanding resources and funding, perhaps due to security concerns (assuming that these machines were produced by China, Russia, or some other world power), but perhaps through a genuine interest in the possibility of these phenomena being of an extra-terrestrial origin. Finally, making matters even more curious, the full version of the report was not even released to the public, raising further questions as to what these objects and entities could possibly be and what might be contained in the final version of the report, which is currently being held from the public at large.